Hölder Continuity of Weak Solutions to Subelliptic Equations with Rough Coefficients

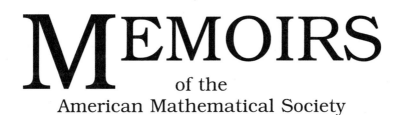

MEMOIRS
of the
American Mathematical Society

Number 847

Hölder Continuity of Weak Solutions to Subelliptic Equations with Rough Coefficients

Eric T. Sawyer
Richard L. Wheeden

March 2006 • Volume 180 • Number 847 (first of 5 numbers) • ISSN 0065-9266

American Mathematical Society
Providence, Rhode Island

2000 *Mathematics Subject Classification.*
Primary 35B65, 35D10, 35H20.

Library of Congress Cataloging-in-Publication Data

Sawyer, E. T. (Eric T.), 1951–
 Hölder continuity of weak solutions to subelliptic equations with rough coefficients / Eric T. Sawyer, Richard L. Wheeden.
  p. cm. — (Memoirs of the American Mathematical Society, ISSN 0065-9266 ; no. 847)
 Includes bibliographical references.
 ISBN 0-8218-3826-1 (alk. paper)
 1. Differential equations, Partial—Numerical solutions. 2. Differential equations, Hypoelliptic—Numerical solutions. I. Wheeden, Richard L. II. Title. III. Series.

QA3.A57 no. 847
[QA377]
510 s—dc22
[515′.3533]   2005057158

Memoirs of the American Mathematical Society

This journal is devoted entirely to research in pure and applied mathematics.

Subscription information. The 2006 subscription begins with volume 179 and consists of six mailings, each containing one or more numbers. Subscription prices for 2006 are US$624 list, US$499 institutional member. A late charge of 10% of the subscription price will be imposed on orders received from nonmembers after January 1 of the subscription year. Subscribers outside the United States and India must pay a postage surcharge of US$31; subscribers in India must pay a postage surcharge of US$43. Expedited delivery to destinations in North America US$35; elsewhere US$130. Each number may be ordered separately; *please specify number* when ordering an individual number. For prices and titles of recently released numbers, see the New Publications sections of the *Notices of the American Mathematical Society.*
 Back number information. For back issues see the *AMS Catalog of Publications.*
 Subscriptions and orders should be addressed to the American Mathematical Society, P. O. Box 845904, Boston, MA 02284-5904, USA. *All orders must be accompanied by payment.* Other correspondence should be addressed to 201 Charles Street, Providence, RI 02904-2294, USA.

Memoirs of the American Mathematical Society is published bimonthly (each volume consisting usually of more than one number) by the American Mathematical Society at 201 Charles Street, Providence, RI 02904-2294, USA. Periodicals postage paid at Providence, RI. Postmaster: Send address changes to Memoirs, American Mathematical Society, 201 Charles Street, Providence, RI 02904-2294, USA.

Contents

Abstract

We study interior regularity of weak solutions of second order linear divergence form equations with degenerate ellipticity and rough coefficients. In particular, we show that solutions of large classes of subelliptic equations with bounded measurable coefficients are Hölder continuous. We present two types of results dealing with such equations. The first type generalizes the celebrated Fefferman-Phong geometric characterization of subellipticity in the smooth case. We introduce a notion of L^q-subellipticity for the rough case and develop an axiomatic method which provides a near characterization of the notion of L^q-subellipticity. The second type deals with generalizing a case of Hörmander's celebrated algebraic characterization of subellipticity for sums of squares of real analytic vector fields, namely the case of diagonal vector fields. In this case, we introduce a "flag condition" as a substitute for the Hörmander commutator condition which turns out to be equivalent to it in the real analytic case. The question of regularity for quasilinear subelliptic equations with smooth coefficients provides motivation for our study, and we briefly indicate some applications in this direction, including degenerate Monge-Ampère equations.

Received by the editor 2004/09/23.

2000 *Mathematics Subject Classification*. Primary 35B65, 35D10, 35H20.

Key words and phrases. Regularity, subelliptic equations, rough coefficients, commutator condition, control distance.

Overview

This paper is concerned with regularity of solutions to rough subelliptic equations. Historically, the regularity of weak solutions to second order linear partial differential equations has been reasonably well understood in two cases:

- when the equation is *subelliptic*, and the coefficients are restricted to being *smooth*,
- when the equation is *elliptic*, and the coefficients are quite *rough*.

In the subelliptic case, there are two main types of result. First, there is the algebraic commutator criterion of Hörmander for sums of squares of smooth vector fields. These operators have a special "sum of squares" form for the second order terms, but no additional restriction on the smooth first order term. Second, there is the geometric "control ball" criterion of Fefferman and Phong that applies to operators with general smooth subelliptic second order terms, but the operators are restricted to be self-adjoint.

In the elliptic case, De Giorgi-Nash-Moser theory applies to equations in divergence form with bounded measurable coefficients. The minimal regularity assumptions on the coefficients in this theory make it well suited for applications to elliptic quasilinear equations.

In this paper we initiate a unified treatment of the subelliptic case when the coefficients are also rough, with a view to obtaining a theory sufficiently rich to apply to subelliptic quasilinear equations. Our methods involve a merging of techniques used by De Giorgi-Nash-Moser, Fefferman-Phong, Hörmander and Franchi, and result in an analogue of the Fefferman-Phong theorem for rough coefficients, and an analogue of the Hörmander theorem for *diagonal* vector fields with rough coefficients.

Much remains to be done in these areas however. For example, there are no results of the Fefferman-Phong type including *general* first order terms, save for the special case of the degenerate heat equation. Moreover, the case of sums of squares of *non-diagonal* subelliptic vector fields with rough coefficients is far from understood except in the simplest subelliptic situations. Finally, the case of *degenerate* subelliptic operators, where the degeneracy is measured by a weight as well as by a subelliptic geometry, leads to theories that do not apply to operators with general lower order terms, and this case has not been pursued here.

CHAPTER 1

Introduction

The point of departure for this work consists of three papers: L. Hörmander [21], C. Fefferman and D. H. Phong [7] and B. Franchi [8]. In [21], Hörmander obtained hypoellipticity, actually subellipticity, of sums of squares of smooth vector fields whose Lie algebra spans at every point. In [7], Fefferman and Phong considered general nonnegative semidefinite smooth linear operators, and characterized subellipticity in terms of a containment condition involving Euclidean balls and "subunit" balls related to the geometry of the nonnegative semidefinite form associated to the operator. The problem of extending these results to include nonlinear operators requires an understanding of subellipticity for linear operators with nonsmooth coefficients, generally as rough as the weak solution. Some nonsmooth degenerate elliptic cases when the eigenvalues all vanish at the same rate were investigated in Fabes, Kenig and Serapioni [6], using A_2 weights or quasiconformal maps and the Euclidean metric to control the degeneracy. The first treatment of a nonsmooth subelliptic case occurred in Franchi and Lanconelli [9], and in greater generality, later in Franchi [8], which we now describe in more detail.

Franchi considered a collection $\mathcal{X} = \{X_j\}_{j=1}^{n}$ of diagonal Lipschitz vector fields $X_j = a_j(x) \frac{\partial}{\partial x_j}$, $1 \leq j \leq n$. Assuming a strong form of a reverse Hölder inequality involving integral curves of vector fields in the span of $\mathcal{X}$, and the subunit balls B of Fefferman and Phong, Franchi established a subrepresentation inequality for Lipschitz functions f similar to, but slightly weaker than,

$$(1) \qquad |f(x) - f_B| \leq C \int_B |\mathcal{X}f(z)| \frac{\delta(x,z)}{|B(x, \delta(x,z))|} dz, \qquad x \in B,$$

where δ is the subunit metric of Fefferman and Phong. From such subrepresentation inequalities, it is possible in many cases to deduce Sobolev and Poincaré inequalities for Lipschitz functions, as well as Harnack inequalities and Hölder continuity of weak solutions to rough equations.

There has recently developed a vast literature dealing with these matters, propelled in large part by the aforementioned works, as well as the pivotal papers of L. Rothschild and E. Stein [36], D. Jerison [22], A. Nagel, E. Stein and S. Wainger [33] and many others. These latter papers deal mainly with smooth vector fields satisfying Hörmander's condition. For example, it was shown in [33] that the underlying metric space is a homogeneous space, and in [22], a Poincaré inequality was derived, with subsequent improvements by G. Lu in [25], [26]. In this paper, we shall be concerned with both the general setting of rough nonnegative semidefinite operators, as well as the special case of operators arising from rough *diagonal* vector fields. Due to the large collection of papers in these areas, we will only mention work with a direct impact on our development, and for a more complete history, we refer the reader to the recent papers [19] and [12] and the references

1

given there, some of which we will discuss in more detail below. See also [24], [27], [28] and references given there, for further results on subrepresentation inequalities. We also refer the reader to [46] for adaptations of the Morrey-Campanato method for rough elliptic sums of smooth vector fields satisfying Hörmander's condition. In this regard see also the limitations of this method discussed at the end of subsection 1.3 below and in appendix A.5 of [34]. Finally, some nonsubelliptic equations are treated in [38], [39], [40] and [35].

We will extend the characterization of subellipticity by Fefferman and Phong in [7] to a near characterization in the rough case (Theorem 8 and Theorem 10), and building on ideas of Franchi in [8], we will use this result to obtain an extension of Hörmander's theorem [21] to operators with quadratic forms comparable to a sum of squares of rough diagonal vector fields (Theorems 12, 17 and 20). The general case of nondiagonal rough vector fields will be considered in a subsequent paper.

We begin by recalling Hörmander's celebrated theorem on hypoellipticity of sums of squares of smooth vector fields ([21]): Suppose

$$X_\ell = \sum_{i=1}^n a_{i\ell}(x) \frac{\partial}{\partial x_i}, \qquad 0 \le \ell \le m,$$

is a collection of smooth vector fields in a domain $\Omega \subset \mathbb{R}^n$ satisfying the following *commutation condition.*

DEFINITION 1. *The vector fields* $\{X_\ell\}_{\ell=0}^m$ *satisfy the* commutation condition *in* Ω *if for every* $x \in \Omega$, *there is* $p \in \mathbb{N}$ *such that the linear span of the vector fields* X_ℓ *and their commutators up to order* p,

$$span\left\{X_{\ell_1}, [X_{\ell_1}, X_{\ell_2}], [X_{\ell_1}, [X_{\ell_2}, X_{\ell_3}]], ..., [X_{\ell_1}, [X_{\ell_2}, [... [X_{\ell_{p-1}}, X_{\ell_p}]...]]] :$$
$$0 \le \ell_j \le m\right\},$$

is equal to $\mathbb{R}^n$.

Then the linear operator

$$L = X_0 + \sum_{\ell=1}^m X_\ell' X_\ell$$

is hypoelliptic in Ω, i.e. a distribution u is smooth in any open set in which Lu is smooth. The commutation condition in Definition 1 is essentially necessary; in particular it holds when the coefficients $a_{i\ell}(x)$ are real-analytic and L is hypoelliptic. The conclusion of Hörmander's theorem can be strengthened to show that L is subelliptic, where in the case $L = \nabla' B(x) \nabla$ is self-adjoint, we say that L is *subelliptic* in Ω if there is $\varepsilon > 0$ such that

$$(2) \qquad \int \text{Re}(uLu) + C\|u\|_{L^2}^2 \ge c\|u\|_{H^\varepsilon}^2,$$

for all $u \in C_c^\infty(\Omega)$, where H^ε denotes the usual Sobolev space of functions with ε derivatives in L^2. A classical result is that subellipticity implies hypoellipticity. Observe that if $X_\ell = \sum_{i=1}^n a_{i\ell}(x) \frac{\partial}{\partial x_i}$, then $\sum_{\ell=1}^m X_\ell' X_\ell = \nabla' B(x) \nabla$ where $B(x) = [b_{ij}(x)]_{i,j=1}^n$, $b_{ij}(x) = \sum_{\ell=1}^m a_{i\ell}(x) a_{j\ell}(x)$. The principal symbol of the operator

$L = X_0 + \sum_{\ell=1}^{m} X_\ell' X_\ell$ is by definition

$$(3) \qquad \sum_{\ell=1}^{m} (X_\ell \cdot \xi)^2 = \sum_{\ell=1}^{m} \left(\sum_{i=1}^{n} a_{i\ell}(x) \xi_i \right)^2 = \xi' B(x) \xi,$$

A characterization of subellipticity for more general, but self-adjoint, smooth operators $L = \nabla' B(x) \nabla$ has been given by Fefferman and Phong in [**7**]. Given a smooth nonnegative semidefinite matrix $B(x)$ defined in Ω, they use the metric

$$(4) \qquad \delta(x,y) = \inf\{r > 0 : \gamma(0) = x, \gamma(r) = y, \gamma \text{ is Lipschitz and subunit}\}$$

on Ω, often referred to as the Carnot-Carathéodory distance, or control distance, where a Lipschitz curve $\gamma : [0,r] \to \Omega$ is *subunit* (with respect to the matrix $B(x)$, or its associated operator L) if

$$(\gamma'(t) \cdot \xi)^2 \le \xi' B(\gamma(t)) \xi, \quad a.e. \ t \in [0,r], \ \xi \in \mathbb{R}^n.$$

Then $L = \nabla' B(x) \nabla$ is subelliptic in Ω if and only if the δ-balls $K(x,r) = \{y \in \Omega : \delta(x,y) < r\}$ satisfy an ε-comparability condition with Euclidean balls $D(x,r) = \{y \in \Omega : |x-y| < r\}$: for every compact subset K of Ω, there are positive constants C, r_0 and ε such that

$$(5) \qquad D(x,r) \subset K(x, Cr^\varepsilon), \quad x \in K, 0 < r \le r_0.$$

We now wish to consider quasilinear equations of the form

$$(6) \qquad \mathcal{L}u \equiv \nabla' B(x, u(x)) \nabla u = f(x, u(x), \nabla u(x)),$$

where $f(x,z,\mathbf{p}) \in C^\infty(\Omega \times \mathbb{R} \times \mathbb{R}^n)$, $B(x,z) = [b_{ij}(x,z)]_{i,j=1}^n \in C^\infty(\Omega \times \mathbb{R})$ and the quadratic form $\xi' B(x, u(x)) \xi$ is nonnegative semidefinite. We will also consider the special case

$$(7) \qquad c \sum_{\ell=1}^{m} (X_\ell(x, u(x)) \cdot \xi)^2 \le \xi' B(x, u(x)) \xi \le C \sum_{\ell=1}^{m} (X_\ell(x, u(x)) \cdot \xi)^2,$$

where the nonlinear vector fields $X_\ell(x, u(x))$ arise from vector fields

$$X_\ell(x,z) = \sum_{i=1}^{n} a_{i\ell}(x,z) \frac{\partial}{\partial x_i}, \quad 1 \le \ell \le m,$$

with coefficients $a_{i\ell}(x,z) \in C^\infty(\Omega \times \mathbb{R})$. In order to obtain regularity results for such equations, it is necessary to treat the linearized operators $\widetilde{L} = \nabla' \widetilde{B}(x) \nabla$ with coefficients $\widetilde{B}(x) = B(x, u(x))$ generally as rough as $u(x)$. In the elliptic case, there exist positive constants c, C such that

$$c |\xi|^2 \le \xi' B(x, u(x)) \xi \le C |\xi|^2$$

(in Definition 1, this occurs when $p = 1$; the vector fields $\{X_\ell\}_{\ell=1}^m$ then span $\mathbb{R}^n$ at each point of Ω), and then the results of De Giorgi - Nash - Moser theory can be applied (see e.g. [**30**]): If

$$L = \nabla' B(x) \nabla = \sum_{i,j=1}^{n} \frac{\partial}{\partial x_i} b_{ij}(x) \frac{\partial}{\partial x_j}$$

where the coefficients $b_{ij}(x)$ are bounded and measurable in Ω and satisfy the ellipticity condition

$$c |\xi|^2 \le \xi' B(x) \xi \le C |\xi|^2, \quad a.e. \ x \in \Omega, \xi \in \mathbb{R}^n,$$

then all weak $W^{1,2}(\Omega)$ solutions u to the equation $Lu = 0$ are Hölder continuous in Ω. Here $W^{1,2}(\Omega)$ is the Sobolev space of square integrable functions f on Ω with square integrable gradients ∇f. A bounded function f is Hölder continuous of order α on Ω if $|f(x) - f(y)| \leq C|x - y|^\alpha$ for $x, y \in \Omega$.

However, in the nonelliptic setting where $\widetilde{B}(x)$ is nonnegative semidefinite (in Definition 1, the commutation condition fails with $p = 1$), we must consider linear operators $\widetilde{L} = \nabla' \widetilde{B}(x) \nabla$ with bounded measurable coefficients in Ω and degenerate nonnegative semidefinite quadratic forms, in particular those satisfying

$$c \sum_{\ell=1}^{m} \left(\widetilde{X}_\ell(x) \cdot \xi\right)^2 \leq \xi' \widetilde{B}(x) \xi \leq C \sum_{\ell=1}^{m} \left(\widetilde{X}_\ell(x) \cdot \xi\right)^2, \quad a.e.\ x \in \Omega,$$

where $\widetilde{X}_\ell(x) = X_\ell(x, u(x)) = [a_{i\ell}(x, u(x))]_{i=1}^n$. We obtain regularity results in both the general and particular settings mentioned above.

First, we obtain an analogue of the theorem of Fefferman and Phong that is valid for rough operators $L = \nabla' B(x) \nabla$, with a general nonnegative semidefinite bounded measurable matrix $B(x)$. Before describing our result, we point out that an example in Xu [45] shows that the ε in (5) cannot be taken equal to the ε in (2) if $B(x)$ is not sufficiently differentiable (at least $C^{\frac{1}{2}(\frac{1}{\varepsilon}-1)}$ is needed). Xu also showed that in two dimensions, (5) implies (2) with a smaller $\varepsilon > 0$, if $B(x)$ is twice continuously differentiable. Note however that the notion of subellipticity of order ε in (2) is based on L^2-Sobolev spaces, while the notion of L^q-subellipticity we use in this paper (see Definition 6 below) is based on Hölder spaces. Roughly speaking, we show that weak solutions of equations involving L are Hölder continuous for some positive exponent provided the ε-comparability condition (5) holds, there is a doubling condition on the balls $K(x, r)$, and certain Poincaré and Sobolev inequalities hold relating the subunit metric to the degenerate form $\xi' B(x) \xi$. C. Gutiérrez and E. Lanconnelli [19] have obtained a special case of this result, which we discuss following Theorem 10 below. The ε-comparability condition (5) often turns out to be necessary in this rough setting, and the Sobolev and Poincaré inequalities are essentially necessary for certain related notions of subellipticity for Dirichlet and Neumann boundary value problems. We also give a version for more general quasimetrics, in place of the subunit metric, which will see application to regularity of weak solutions for operators L with principal symbol satisfying (7).

Second, we obtain an analogue of Hörmander's theorem for rough vector fields $\{X_\ell\}_{\ell=1}^m$ in the special case where the linear term is absent, so that the operator is self-adjoint, and the vector fields are diagonal, i.e. of the form

$$\{X_\ell\}_{\ell=1}^m = \left\{ a_j(x) \frac{\partial}{\partial x_j} \right\}_{j=1}^n,$$

with $m = n$. We introduce an analogue of Hörmander's commutation condition in this setting, called the flag condition, which requires that for each $x \in \Omega$ and each index set $\phi \subset \mathcal{I} \subsetneq \{1, 2, ..., n\}$, there is $j \notin \mathcal{I}$ such that a_j does not vanish identically on $(x + \mathcal{V}_\mathcal{I}) \cap \mathcal{N}$ for any neighbourhood $\mathcal{N}$ of x in Ω, where $\mathcal{V}_\phi = \{0\}$ and $\mathcal{V}_\mathcal{I} = span\{\mathbf{e}_i : i \in \mathcal{I}\}$. Roughly speaking, we show that weak solutions of an operator $L = \nabla' B(x) \nabla$, whose principal symbol is equivalent to the quadratic form $\sum_{j=1}^n (a_j(x) \xi_j)^2$, are Hölder continuous for some positive exponent provided the flag condition holds, along with certain reverse Hölder conditions on the coefficients of the vector fields. The flag condition turns out to be necessary in this rough setting, and is in fact equivalent to Hörmander's commutation condition when the vector fields are analytic. We now consider these theorems in more detail.

We consider operators of the form $L = \nabla' B(x) \nabla = \sum_{i,j=1}^{n} \frac{\partial}{\partial x_i} b_{ij}(x) \frac{\partial}{\partial x_j}$ where the real-valued coefficients $b_{ij}(x)$ are bounded and measurable in Ω and the quadratic form $\xi' B(x) \xi$ is nonnegative semidefinite. We seek conditions on a nonnegative semidefinite quadratic form

$$\mathcal{Q}(x, \xi) = \xi' Q(x) \xi,$$

in order that all operators L with principal symbol $\xi' B(x) \xi$ equivalent to $\mathcal{Q}(x, \xi)$ are *subelliptic* in Ω. To explain this more precisely, we recall the following two definitions (see Fefferman and Phong [**7**] and Guan [**15**]). A vector field $T = \sum_{i=1}^{n} \alpha_i(x) \frac{\partial}{\partial x_i}$, with bounded coefficients α_i, is *subunit* with respect to a symmetric nonnegative matrix $B(x)$ in Ω if

$$\left(\sum_{i=1}^{n} \alpha_i(x) \xi_i \right)^2 \leq \xi' B(x) \xi, \qquad a.e. \ x \in \Omega, \xi \in \mathbb{R}^n.$$

DEFINITION 2. *Let $0 < \alpha < 1$. A linear operator*

$$L = \nabla' B(x) \nabla = \sum_{i,j=1}^{n} \frac{\partial}{\partial x_i} b_{ij}(x) \frac{\partial}{\partial x_j}$$

is α-subelliptic in Ω if there is a positive function $\mathcal{C}(E, z_1, z_2, z_3, z_4)$ defined on $\mathcal{P}(\Omega) \times [0, \infty)^4$ (where $\mathcal{P}(\Omega)$ is the lattice of compact subsets of Ω), increasing in each variable separately, such that for all N-tuples $\mathbf{T} = (T_1, ..., T_N)$ of bounded subunit (with respect to $B(x)$) vector fields, all bounded functions $f, \mathbf{g}$, and all compact subsets K of Ω, every weak solution $u \in W^{1,2}(\Omega)$ to the divergence form equation

$$Lu = f + \mathbf{T}' \mathbf{g},$$

satisfies, possibly after redefinition on a set of measure zero,

$$\|u\|_{C^\alpha(K)} \leq \mathcal{C},$$

where $\mathcal{C} = \mathcal{C}(K, \|u\|_2, \|f\|_\infty, \|\mathbf{g}\|_\infty, N)$. Here $\mathbf{T}'$ denotes the transpose of $\mathbf{T}$ and

$$(8) \qquad \|w\|_{C^\alpha(K)} = \sup_{x \in K} |w(x)| + \sup_{x,y \in K} \frac{|w(x) - w(y)|}{|x - y|^\alpha}.$$

Note that the Hölder exponent α measures the gain in smoothness of the weak solution u ($\in C^\alpha$) over the data f and $\mathbf{g}$ ($\in L^\infty \sim C^0$). This gain will clearly depend on the operator L, but will be independent of the norms $\|u\|_2$, $\|f\|_\infty$ and $\|\mathbf{g}\|_\infty$ by the linearity of the equation: if $\widetilde{u} = cu$, $\widetilde{f} = cf$ and $\widetilde{\mathbf{g}} = c\mathbf{g}$ for a nonzero real number c, then u is a solution to $Lu = f + \mathbf{T}' \mathbf{g}$ if and only if $\widetilde{u}$ is a solution to $L\widetilde{u} = \widetilde{f} + \mathbf{T}' \widetilde{\mathbf{g}}$. The norms $\|\widetilde{u}\|_2$, $\left\|\widetilde{f}\right\|_\infty$ and $\|\widetilde{\mathbf{g}}\|_\infty$ can thus be made arbitrarily small while the Hölder exponent α of $\widetilde{u}$ remains unchanged.

REMARK 3. *We caution the reader that in subsequent subsections we will strengthen the above notion of subellipticity for rough linear operators L by requiring the same Hölder conclusion, but for more general equations with lower order terms and rougher data. The Hölder exponent α will then depend on the degree of roughness of the data as well as the norms of the coefficients of the lower order terms, and this will have to be reflected in the definition of subellipticity. Moreover,* *as the data f and $\mathbf{g}$ will be permitted to lie in rougher Lebesgue spaces $L^{\frac{q}{2}}$ and L^q respectively, the quantity α will no longer reflect the gain in smoothness of u over the data f and $\mathbf{g}$. Instead, the gain from f and $\mathbf{g}$ will be measured by the quantities $\frac{2n}{q}$ and $\frac{n}{q}$ respectively, where q is the optimal exponent such that weak solutions u are Hölder continuous of some positive order for data f and $\mathbf{g}$ in $L^{\frac{q}{2}}$ and L^q respectively. Recall that classical fractional integration of order α maps L^q to C^ε if $\varepsilon = \alpha - \frac{n}{q} > 0$.*

As a result of these considerations, we will no longer specify the Hölder exponent α as part of the definition of subellipticity of L, but will instead specify the Lebesgue exponent q of the data, resulting in a definition of L^q-subellipticity for the operator L in Definition 6 below. We will also define in Definition 7 a notion of L^q-subellipticity for a nonnegative semidefinite quadratic form $\mathcal{Q}(x,\xi) = \xi' Q(x) \xi$, which will essentially turn out to be that every rough linear operator L, even with lower order terms, whose principal symbol is comparable to $\mathcal{Q}(x,\xi)$, is L^q-subelliptic. Finally in Definition 11, we will define a collection of vector fields $\{X_i(x)\}_{i=1}^m$ to be L^q-subelliptic if the quadratic form $\mathcal{Q}(x,\xi) = \sum_{i=1}^m (X_i(x) \cdot \xi)^2$ associated to the operator $L = \sum_{i=1}^m X_i(x)' X_i(x)$ is L^q-subelliptic. In all of these cases, we will refer to L^∞-subelliptic, the weakest of the L^q-subelliptic definitions, as simply subelliptic.

1. An extension of the Fefferman-Phong theorem to rough operators

We extend the definition of the subunit metric $\delta(x,y)$ to the case of a continuous quadratic form $\mathcal{Q}(x,\xi)$.

DEFINITION 4. *If $\mathcal{Q}(x,\xi)$ is a continuous quadratic form on Ω, we define*

$$\delta(x,y) = \inf\{r > 0 : \gamma(0) = x, \gamma(r) = y, \gamma \text{ is Lipschitz and subunit in } \Omega\},$$

where a Lipschitz curve $\gamma : [0,r] \to \Omega$ is subunit with respect to $\mathcal{Q}(x,\xi)$ if

$$(\gamma'(t) \cdot \xi)^2 \leq \mathcal{Q}(\gamma(t),\xi), \quad a.e.\ t \in [0,r],\ \xi \in \mathbb{R}^n.$$

The function $\delta : \Omega \times \Omega \to [0,\infty]$ is a symmetric metric on Ω since the family of Lipschitz subunit curves in Ω is closed under concatenation and invariant under time reversal. If δ is finite on $\Omega \times \Omega$ we define the *subunit balls* $K(x,r)$ by

$$K(x,r) = \{y \in \Omega : \delta(x,y) < r\}, \quad x \in \Omega,\ 0 < r < \infty.$$

We first consider a general subellipticity theorem that is an analogue of the Fefferman-Phong theorem for rough operators, but involves a more general quasimetric in place of the metric of Fefferman and Phong. In order to state our general theorem, we introduce some notation. A quasimetric d on $\Omega \subset \mathbb{R}^n$ is a finite nonnegative function on $\Omega \times \Omega$ satisfying

$$\begin{aligned} d(x,y) &= 0 \Longleftrightarrow x = y \\ d(x,y) &\leq \kappa(d(x,z) + d(y,z)) \end{aligned}$$

for all x,y,z in Ω. The quasimetric balls $B(x,r)$ are defined by

$$B(x,r) = \{y \in \Omega : d(x,y) < r\}, \quad 0 < r < \infty.$$

Provided the quasimetric $d(x,y)$ is Lebesgue measurable in the second variable, the upper and lower dimensions, Q^* and Q_*, of a quasimetric space with balls $B(x,r)$

are given by

$$(9) \qquad Q^* = \limsup_{r \to 0} \max_{x \in \Omega} \frac{\log |B(x,r)|}{\log r},$$

$$Q_* = \liminf_{r \to 0} \min_{x \in \Omega} \frac{\log |B(x,r)|}{\log r}.$$

Provided the balls $B(x,r)$ are contained in Euclidean balls $D(x, C_{euc}r)$ for some fixed positive constant C_{euc}, which will be the case for all balls considered in this paper, we have $Q^* \geq Q_* \geq n$. Note that for any $q^* > Q^*$ and $q_* < Q_*$, we have the estimates

$$(10) \qquad c_{q^*} r^{q^*} \leq |B(x,r)| \leq C_{q_*} r^{q_*}, \qquad 0 < r < 1, x \in \Omega,$$

for some $c_{q^*}, C_{q_*} > 0$, which explains the terminology.

> **Convention:** Throughout the paper we will have occasion to consider local properties of balls $B(x,r)$ that require x and r to be suitably restricted. We will effect these restrictions by taking a sufficiently small positive constant δ and qualifying our local properties by
>
> $$x \in \Omega, 0 < r < \delta \, dist(x, \partial \Omega),$$
>
> where $dist(x, E)$ denotes the Euclidean distance from the point x to the set E. The positive number δ may change from line to line, but we will only indicate this by a different symbol, such as δ_0 or δ', if a comparison is necesssary.

We will require the following containment, which for the subunit balls $K(x,r)$ is essentially necessary for the notion of subellipticity of the form $\mathcal{Q}$ that is given in Definition 7 below: there are positive constants C, ε and δ such that

$$(11) \qquad D(x,r) \subset B(x, Cr^\varepsilon), \qquad x \in \Omega, 0 < r < \delta \, dist(x, \partial \Omega).$$

This shows that $|B(x,r)| \geq \left| D\left(x, \left(\frac{r}{C}\right)^{\frac{1}{\varepsilon}}\right) \right| = cr^{\frac{n}{\varepsilon}}$ for small r, and hence that $Q^* \leq \frac{n}{\varepsilon}$ is finite whenever (11) holds. There is the following partial converse.

REMARK 5. *If the balls $B(x,r)$ are locally equivalent to their convex hulls $coB(x,r)$, i.e.*

$$(12) \qquad coB(x,r) \subset B(x, Cr), \qquad x \in \Omega, 0 < r < \delta dist(x, \partial \Omega),$$

for some $C, \delta > 0$, and uniformly bounded for $x \in \Omega$, $0 < r < \delta dist(x, \partial \Omega)$, then the containment condition (11) is a consequence of the finiteness of the upper dimension Q^* (which in turn is a consequence of the doubling condition (13) below). Indeed, there is a positive definite matrix A with corresponding ellipsoids

$$E(\overline{x}, t) = \{y \in \mathbb{R}^n : \|A(\overline{x} - y)\| < t\}, \qquad t > 0,$$

centred at $\overline{x}$, the center of mass of $coB(x,r)$, such that

$$E(\overline{x}, r) \subset coB(x,r) \subset E\left(\overline{x}, n^{\frac{3}{2}} r\right).$$

This variant of F. John's famous result can be found in Gutiérrez ([18] Theorem 1.8.2). If Λ is the largest eigenvalue of A and $\rho = \Lambda^{-1} r$, then $D(\overline{x}, \rho) \subset E(x,r)$. If C_b is a bound for the diameters of the balls $B(x, Cr)$ with $x \in \Omega, 0 < r \leq 1$, then $\sup_{x\in\Omega,0<r\leq1} diam\{E(\bar{x},r)\}\leq C_b$, and so there is a constant C_b' such that $\lambda^{-1}r\leq C_b'$ for any eigenvalue λ of A. Thus we have

$$|coB(x,r)|\leq\left|E\left(\bar{x},n^{\frac{3}{2}}r\right)\right|=C_nr^n(\det A)^{-1}\leq C_n'(C_b')^{n-1}\rho.$$

Now (10) yields $c_{q^*}r^{q^*}\leq|B(x,r)|$ for $q^*>Q^*$ and some $c_{q^*}>0$, and altogether we have

$$c_{q^*}r^{q^*}\leq|B(x,r)|\leq|coB(x,r)|\leq C_n'(C_b')^{n-1}\rho.$$

Thus the convex hull $coB(x,r)$ of $B(x,r)$ contains a Euclidean ball $D(x,cr^{q^*})$ with $c=\frac{c_{q^*}}{C_n'(C_b')^{n-1}}$, and then by (12), $D\left(x,c\left(\frac{r}{C}\right)^{q^*}\right)\subset coB\left(x,\frac{r}{C}\right)\subset B(x,r)$.

We will also require the doubling condition

$$(13)\qquad |B(x,2r)|\leq C|B(x,r)|,\qquad x\in\Omega,0<r<\infty,$$

which makes $(\Omega,d,|\cdot|)$ into a general homogeneous space. See the beginning of subsection 2.2 for a detailed discussion of such spaces. Note that doubling (13) implies (18) below for some finite exponent D, which yields $Q^*\leq D<\infty$.

We now introduce a bounded nonnegative semidefinite quadratic form $\mathcal{Q}(x,\xi)=\xi'Q(x)\xi$, where $Q(x)$ is a symmetric matrix for each $x\in\Omega$, and define

$$(14)\qquad \|\mathbf{U}(x)\|_\mathcal{Q}^2=\mathcal{Q}(x,\mathbf{U}(x))=\mathbf{U}(x)'Q(x)\mathbf{U}(x)$$

for any vector-valued function $\mathbf{U}(x)$. The Sobolev inequality we need is: there is $\sigma>1$ and $\delta>0$ such that for all balls $B=B(y,r)$ with $y\in\Omega$, $0<r<\delta\,dist(y,\partial\Omega)$,

$$(15)\qquad \left\{\frac{1}{|B|}\int_B|w|^{2\sigma}\right\}^{\frac{1}{2\sigma}}\leq Cr\left\{\frac{1}{|B|}\int_B\|\nabla w\|_\mathcal{Q}^2\right\}^{\frac{1}{2}}+C\left\{\frac{1}{|B|}\int_B|w|^2\right\}^{\frac{1}{2}},$$

for all $w\in W_0^{1,2}(B)$. In many applications the stronger form of hypothesis (15) holds without the L^2 average of w on the far right. Note that the right-hand side of (15) is comparable to the normalized $\mathcal{Q}$-Sobolev norm

$$(16)\qquad \|w\|_{W_\mathcal{Q}^{1,2}(B)}^*\equiv\left\{\frac{1}{|B|}\int_B\left(\|r\nabla w\|_\mathcal{Q}^2+|w|^2\right)\right\}^{\frac{1}{2}}.$$

Since $\|w\|_{W_\mathcal{Q}^{1,2}(B)}^*$ is at most $\sqrt{\|Q\|_\infty}$ times the classical normalized Sobolev norm $\|w\|_{W^{1,2}(B)}^*$, we see that $2\sigma\leq\frac{2n}{n-2}$. The Poincaré inequality we need is: there is $C_0\geq1$ and $\delta>0$ such that for all balls $B=B(y,r)$ and $B^*=B(y,C_0r)$ with $y\in\Omega$, $0<C_0r<\delta\,dist(y,\partial\Omega)$,

$$(17)\qquad \left\{\frac{1}{|B|}\int_B\left|w-\left(\frac{1}{|B|}\int_Bw\right)\right|^2\right\}^{\frac{1}{2}}\leq Cr\left\{\frac{1}{|B^*|}\int_{B^*}\|\nabla w\|_\mathcal{Q}^2\right\}^{\frac{1}{2}},$$

for every $w\in W^{1,2}(B^*)$. In applications to vector fields $\mathcal{X}=\{X_j\}_{j=1}^n$, the Sobolev and Poincaré inequalities (15) and (17) are typically deduced from a subrepresentation formula like (1), see e.g. Proposition 74 below, and hence the stronger form of hypothesis (15) holds without the L^2 average of w on the far right. Sharper versions of (17) are also often available in special cases; see Proposition 74 regarding different exponents, and see Remark 81 in section 4.4.1 regarding the choice $B^*=B$. See Proposition 44 for yet more on subrepresentation formulas.

We remark that in the case of the Euclidean metric, we have $\mathcal{Q}(x,\xi) = |\xi|^2$ and that (15) holds for $\sigma = \frac{n}{n-2} = \left(\frac{n}{2}\right)'$, $n \geq 3$. Typically $\sigma = \frac{D}{D-2}$ where D is the doubling exponent for the homogeneous space with balls $B(x,r)$, given in the inequality

(18) $$|B(x,r)| \leq C\left(\frac{r}{t}\right)^D |B(y,t)|, \qquad B(x,r) \supset B(y,t).$$

See Lemma 73 in subsection 4.2. The following quantity will play the role of "dimension" in the sequel:

$$Q = \max\{Q^*, 2\sigma'\},$$

where $\frac{1}{\sigma} + \frac{1}{\sigma'} = 1$. Note that $2\sigma' = D$ if $\sigma = \frac{D}{D-2}$, and that $D \geq Q^*$, where Q^* is the upper dimension of the homogeneous space. Thus in the typical case where $\sigma = \frac{D}{D-2}$, we have $Q = D$.

The next hypothesis is crucial for Moser iteration, and as we show in Proposition 68 below, holds automatically with $p = \infty$ for the subunit balls $K(x,r)$, if $\mathcal{Q}(x,\xi)$ is continuous in x and (11) holds for the subunit balls:

(19) $$D(x,r) \subset K(x, Cr^\varepsilon), \qquad x \in \Omega, 0 < r < \delta\, dist(x, \partial\Omega).$$

Similar results for subunit balls, but with δ-Lipschitz cutoff functions instead of ordinary Lipschitz cutoff functions, have been obtained earlier in [12] and [11] and will be discussed below. We suppose there are positive constants c, N and δ such that for each ball $B(y,r)$ with $y \in \Omega$, $0 < r < \delta\, dist(y, \partial\Omega)$, there is an accumulating sequence of Lipschitz cutoff functions $\{\psi_j\}_{j=1}^\infty$ on $B(y,r)$ with the following five properties ($E \Subset F$ means that the closure of E is contained in the interior of F):

(20) $$\begin{cases} supp\,\psi_1 & \subset\ B(y,r), \\ B(y,cr) & \subset\ \{x: \psi_j(x) = 1\}, \quad j \geq 1 \\ supp\,\psi_{j+1} & \Subset\ \{x: \psi_j(x) = 1\}, \quad j \geq 1 \\ \psi_j & is\ Lipschitz, \quad j \geq 1 \\ \left\{\frac{1}{|B(y,r)|}\int \|\nabla\psi_j\|_\mathcal{Q}^p\,dx\right\}^{\frac{1}{p}} & \leq\ C_p \frac{j^N}{r}, \quad j \geq 1 \end{cases}$$

for some $p > 2\sigma'$. Note that $2\sigma' \geq n$ since $2\sigma \leq \frac{2n}{n-2}$, as we observed following (15). Thus the condition postulates an accumulating sequence $\{\psi_j\}_{j=1}^\infty$ of Lipschitz cutoff functions with L^p control of $\|\nabla\psi_j\|_\mathcal{Q}$ for an exponent $p > 2\sigma' \geq n$, where $\sigma > 1$ is the gain in the Sobolev inequality (15), uniformly for $B(y,r)$ with $y \in \Omega$, $0 < r < \delta\, dist(y, \partial\Omega)$.

In the case that the balls $B(y,r)$ are the subunit balls $K(y,r)$, the containment condition (19) holds and the quadratic form $\mathcal{Q}(x,\xi)$ is continuous in x, then the "accumulating sequence of Lipschitz cutoff functions" condition (20) holds automatically with $p = \infty$. Indeed, this follows formally from the following fundamental inequality relating the subunit metric δ to the corresponding quadratic form $\mathcal{Q}$:

(21) $$\|\nabla_x \delta(x,y)\|_\mathcal{Q} \leq \sqrt{n} + C,$$

where C depends on $\mathcal{Q}$ and the inequality holds in the distribution sense provided $\mathcal{Q}$ is continuous and δ is finite on $\Omega \times \Omega$. Under more restrictive hypotheses, this was first obtained in Garofalo and Nhieu [12] and Franchi, Serapioni and Serra Cassano [11]. Using this, or more precisely a Lipschitz approximation to this inequality when (19) holds, see (176) below, we can employ ordinary cutoff functions φ and compose them with Lipschitz approximations δ^ε to the metric δ to obtain Lipschitz cutoff functions of the form $\psi(x) = \varphi(\delta^\varepsilon(x,y))$ that satisfy

$$\|\nabla\psi\|_Q \leq \|\varphi'\|_\infty \|\nabla_x\delta^\varepsilon(x,y)\|_Q \leq \sqrt{n}\|\varphi'\|_\infty.$$

See Proposition 68 in section 4 for the details.

Since our methods yield Hölder continuity in more general situations, we will consider the second order equation with lower order terms,

$$(22) \qquad\qquad Lu + \mathbf{HR}u + \mathbf{S}'\mathbf{G}u + Fu = f + \mathbf{T}'\mathbf{g}.$$

The equation now involves a second order term Lu with symbol $\xi'B(x)\xi$, first order terms $\mathbf{HR}u + \mathbf{S}'\mathbf{G}u$ and a zero order term Fu, where $B(x)$ is a bounded measurable nonnegative semidefinite matrix, $\mathbf{R} = \{R_i\}_{i=1}^N$, $\mathbf{S} = \{S_i\}_{i=1}^N$ and $\mathbf{T} = \{T_i\}_{i=1}^N$ are collections of vector fields subunit with respect to $B(x)$; and the operator coefficients $\mathbf{H} = \{H_i\}_{i=1}^N$, $\mathbf{G} = \{G_i\}_{i=1}^N$, F, and the inhomogeneous data $\mathbf{g} = \{g_i\}_{i=1}^N$ and f are measurable. Our hypothesis on the operator coefficients $\mathbf{H}$, $\mathbf{G}$ and F is

$$(23) \qquad\qquad \|F\|_{L^{\frac{q}{2}}(\Omega)} + \|\mathbf{G}\|_{L^q(\Omega)} + \|\mathbf{H}\|_{L^q(\Omega)} \equiv N_q < \infty,$$

for some $q > Q$. Our hypothesis on the inhomogeneous data $f, \mathbf{g}$ is

$$(24) \qquad\qquad \|f\|_{L^{\frac{q}{2}}(\Omega)} + \|\mathbf{g}\|_{L^q(\Omega)} \equiv N_q' < \infty,$$

for the same $q > Q$ as in (23), but with N_q' not necessarily the same as N_q in (23). The distinction between N_q and N_q' is made here because the Hölder continuity exponent α of weak solutions u to (22) will turn out to depend on the gaps $1 - \frac{Q}{p}$ and $1 - \frac{Q}{q}$, where p is as in (20) and q is as in (23) and (24), as well as C_p in (20) and N_q in (23), but *not* on u or N_q' in (24). This is of paramount importance in applications to nonlinear equations such as (51) - see also [**34**].

We now wish to capture in a precise way the notion that a bounded nonnegative semidefinite quadratic form $\mathcal{Q}(x,\xi)$ is *subelliptic* if all operators of the above form, with principal symbol comparable to $\mathcal{Q}(x,\xi)$, are subelliptic operators in the spirit of Definition 2. First we recall that u is a weak solution of (22) in Ω if $u \in W^{1,2}_{loc}(\Omega)$ and satisfies

$$-\int(\nabla u)'\, B\nabla w + \int(\mathbf{HR}u)w + \int u\mathbf{GS}w + \int Fuw = \int fw + \int \mathbf{g}\mathbf{T}w,$$

for all nonnegative $w \in W^{1,2}_0(\Omega)$ (equivalently, we could test over all $w \in W^{1,2}_0(\Omega)$). Note that $\nabla' = div$, and that the prime in $(\nabla u)'$ refers to transpose of a vector rather than adjoint of an operator. By the Sobolev embedding theorem $W^{1,2}_{loc}(\Omega) \subset L^{q_n}_{loc}(\Omega)$ ($q_n = \frac{2n}{n-2}$ if $n \geq 3$ and $q_n < \infty$ if $n = 2$), the individual integrals above converge absolutely if (24) and (23) hold with $q \geq n$ (recall that if the balls $B(x,r)$ are contained in Euclidean balls $D(x,Cr)$, then $Q^* \geq Q_* \geq n$).

For a discussion of alternate definitions of weak solution, and why we focus on the classical definition in this paper, see subsection 6.7 of the appendix. We now strengthen the notion of subelliptic operator given in Definition 2.

DEFINITION 6. *Let $q \in [2,\infty]$. We say that an **operator** $L = \nabla'B(x)\nabla$ with bounded measurable matrix $B(x)$ is L^q-subelliptic in Ω if there are positive functions $\alpha = \alpha(E, z_1)$ and $\mathcal{C} = \mathcal{C}(E, z_1, z_2, z_3)$ defined on $\mathcal{P}(\Omega) \times [0,\infty)$ and $\mathcal{P}(\Omega) \times [0,\infty)^3$ respectively, increasing in each variable separately, such that* every *weak solution u of (22) in Ω satisfies, possibly after redefinition on a set of measure zero,*

(25)
$$\|u\|_{C^{\alpha}(K)} \leq \mathcal{C},$$

for

(26)
$$\alpha = \alpha\left(K, N_q\right),$$
$$\mathcal{C} = \mathcal{C}\left(K, N_q, N_q', \|u\|_2\right),$$

whenever K is a compact subset of Ω, (24) and (23) hold, and $\mathbf{R} = \{R_i\}_{i=1}^{N}$, $\mathbf{S} = \{S_i\}_{i=1}^{N}$ and $\mathbf{T} = \{T_i\}_{i=1}^{N}$ are collections of vector fields subunit with respect to $B(x)$. We say that the operator L is subelliptic in Ω if it is L^{∞}-subelliptic in Ω.

DEFINITION 7. *Let $q \in [2, \infty]$. We say that a bounded measurable nonnegative semidefinite **quadratic form** $\mathcal{Q}(x, \xi)$ is L^q-subelliptic in Ω if every operator $L = \nabla' B(x) \nabla$ whose matrix $B(x)$ satisfies*

(27)
$$c_{sym}\mathcal{Q}(x, \xi) \leq \xi' B(x) \xi \leq C_{sym}\mathcal{Q}(x, \xi), \qquad a.e.\ x \in \Omega, \xi \in \mathbb{R}^n,$$

*for positive constants c_{sym} and C_{sym}, is L^q-subelliptic in Ω, **and** provided the positive functions α and $\mathcal{C}$ in (26) can be chosen to depend only on the constants c_{sym} and C_{sym} in (27) and not on L itself, i.e.*

$$\alpha = \alpha_{c_{sym},C_{sym}}\left(K, N_q\right),$$
$$\mathcal{C} = \mathcal{C}_{c_{sym},C_{sym}}\left(K, N_q, N_q', \|u\|_2\right).$$

We say that the form $\mathcal{Q}(x, \xi)$ is subelliptic in Ω if it is L^{∞}-subelliptic in Ω.

THEOREM 8. *Suppose that $\mathcal{Q}(x, \xi)$ is a bounded measurable nonnegative semidefinite quadratic form in Ω. Let $d(x, y)$ be a symmetric quasimetric in Ω with $d(x, y) \geq c|x - y|$ for some $c > 0$, that is Lebesgue measurable in each variable separately, with upper dimension Q^*, and suppose $\sigma > 1$. Then $\mathcal{Q}(x, \xi)$ is L^q-subelliptic in Ω for*

$$q > Q \equiv \max\{Q^*, 2\sigma'\}, \qquad \frac{1}{\sigma} + \frac{1}{\sigma'} = 1,$$

provided that the following hold for the d-balls $B(x, r) = \{y \in \Omega : d(x, y) < r\}$:

(1) *the doubling condition (13) holds,*
(2) *the containment condition (11) holds,*
(3) *the Sobolev and Poincaré inequalities (15) and (17) hold with the given σ,*
(4) *the "accumulating sequence of Lipschitz cutoff functions" condition (20) holds for some $p > \max\{2\sigma', 4\}$.*

Of course the functions $\alpha_{c_{sym},C_{sym}}$ and $\mathcal{C}_{c_{sym},C_{sym}}$ in Definition (7) will depend on the various constants in the conditions (13), (11), (15), (17) and (20). The proof of the theorem relies on the Moser iteration method, using the quasimetric $d(x, y)$ in place of the Euclidean metric $|x - y|$ and is presented in section 3. See Remark 58 in section 3.1 for a discussion of the restriction $p > 4$ in the fourth condition of Theorem 8.

REMARK 9. *If the containment condition (11) is not required in Theorem 8, then we still obtain Hölder continuity of weak solutions, but with the Euclidean metric replaced by the quasimetric d in the definition of $C^\alpha(K)$:*

$$(28) \qquad \|w\|_{C^\alpha_{quasi}(K)} = \sup_{x \in K} |w(x)| + \sup_{x,y \in K} \frac{|w(x) - w(y)|}{d(x,y)^\alpha}.$$

This is discussed further in the appendix.

A natural family of balls to consider in connection with Theorem 8 is the family of balls $K(x,r)$ arising from the subunit metric $\delta(x,y)$ associated to a quadratic form $\mathcal{Q}(x,\xi) = \xi'Q(x)\xi$ as in (4) with $B(x)$ replaced by $Q(x)$, provided $\delta(x,y)$ is finite on $\Omega \times \Omega$. These balls are measurable by Lemma 66 in subsection 4.2. By Proposition 68, the "accumulating sequence of Lipschitz cutoff functions" condition (20) is automatic with $p = \infty$ for the subunit balls $K(x,r)$ in case the form $\mathcal{Q}(x,\xi)$ is continuous in x and the containment condition (19) holds. Thus we have as a corollary the following theorem that extends the Fefferman-Phong theorem, in that it comes close to characterizing subellipticity in the rough setting. We remark that while the subunit metric is well-defined by (4) even in the case when $\mathcal{Q}(x,\xi)$ is only a bounded Borel measurable function of x, it may exhibit pathological properties if continuity is violated. See Example 48 in section 2.3.1.

THEOREM 10. *Suppose that $\mathcal{Q}(x,\xi)$ is a nonnegative semidefinite continuous quadratic form in Ω, and suppose that the subunit metric $\delta(x,y)$ is finite on $\Omega \times \Omega$. Let the corresponding subunit balls $K(x,r)$ have upper dimension Q^*, and suppose that $\sigma > 1$. Then $\mathcal{Q}(x,\xi)$ is L^q-subelliptic in Ω for $q > Q = \max\{Q^*, 2\sigma'\}$ provided that:*

1. *the doubling condition $|K(x,2r)| \leq C |K(x,r)|$ holds for $0 < r < \infty$,*
2. *the containment condition (19) holds,*
3. *the Sobolev and Poincaré inequalities (15) and (17) hold with $B(x,r) = K(x,r)$.*

We mention here a special case of Theorem 10 that arises in the work of Gutiérrez and Lanconelli [19] on quadratic forms arising from rough dilation invariant vector fields. Assuming a doubling condition and a strong Poincaré inequality relative to a collection of dilation invariant Lipschitz continuous vector fields $\{X_j\}_{j=1}^m$, they prove a Harnack inequality for weak solutions to (22) with $\mathbf{H}$, $\mathbf{G}$, F and $\mathbf{g}$ identically zero, and where the symbol of L is comparable to $\sum_{j=1}^m (X_j \cdot \xi)^2$. See the references there for additional recent work in this general area.

1.1. Almost necessity of the conditions. The second condition in Theorem 10 is necessary, at least in the case that $\mathcal{Q}(x,\xi)$ is continuous and *stably subelliptic* in Ω. We say that a quadratic form $\mathcal{Q}(x,\xi)$ is *stably* subelliptic in Ω if the family of forms

$$\mathcal{Q}_\tau(x,\xi) = \mathcal{Q}(x,\xi) + \tau |\xi|^2, \qquad 0 < \tau < 1,$$

is subelliptic in Ω uniformly in $0 < \tau < 1$, in the sense that the positive functions $\alpha_{C_{sym},C_{sym}}$ and $\mathcal{C}_{C_{sym},C_{sym}}$ in Definition 7 can be chosen independent of τ. Note that $\mathcal{Q}_\tau$ is a small elliptic perturbation of $\mathcal{Q}$, hence the terminology. We note that the conclusions of Theorems 12, 17, 20, 23 and 24 below all persist for stably L^q-subelliptic in place of L^q-subelliptic, since the hypotheses are not compromised upon adding $\tau > 0$ to the coefficients of the vector fields. We do not know if there are quadratic forms $Q(x, \xi)$ that are subelliptic, but not stably subelliptic. The proof that the containment condition (19) is necessary for stable subellipticity is given in Proposition 99 of the appendix. Here, let us merely observe that if we ignore the parameter τ, we can argue formally as follows. If $P(x)$ is a measurable matrix with $P(x)'P(x) = Q(x)$ where $Q(x, \xi) = \xi'Q(x)\xi$, then with y fixed and $u(x) = \delta(x, y)$, the fundamental inequality (21) shows *formally* that

$$Lu = \mathbf{T}'\mathbf{g} = \sum_{j=1}^{n} T_j' g_j'$$

where $L = \nabla'Q(x)\nabla = [P(x)\nabla]'[P(x)\nabla]$, $\mathbf{g} = P(x)\nabla u = (g_j)_{j=1}^n$ has bounded components, and $\mathbf{T} = P(x)\nabla = (T_j)_{j=1}^n$ is a collection of subunit vector fields T_j with respect to $Q(x)$. Indeed, if $P_i(x)$ denotes the ith row of the matrix $P(x)$, then for $1 \le j \le n$,

$$(T_j \cdot \xi)^2 \le \sum_{i=1}^{n} (T_i \cdot \xi)^2 = \sum_{i=1}^{n} (P_i(x) \cdot \xi)^2 = |P(x)\xi|^2$$
$$= (P(x)\xi)' P(x)\xi = \xi'P(x)'P(x)\xi = \xi'Q(x)\xi.$$

The subellipticity of $Q(x, \xi)$ then shows that u is Hölder continuous of order $\alpha > 0$, and it follows that

$$\delta(x, y) = |u(x) - u(y)| \le C |x - y|^\alpha,$$

which is (19) with $\varepsilon = \alpha$. This proof breaks down since (21) is not generally true in the ordinary $W^{1,2}$ sense, and Proposition 99 establishes the necessity of (19) for stable subellipticity by using the Lipschitz approximation (176) in place of (21). We note however, that for the alternate notion of weak solution discussed in subsection 6.5 of the appendix, the distributional inequality (21) of [**12**] and [**11**] often suffices. In certain cases when Q is a sum of squares, such as in the next section, we can show the necessity of the containment condition (19) for mere subellipticity, rather than stable subellipticity, of Q (see Proposition 100 in the appendix).

We turn now to the third condition in Theorem 10. We are unable to demonstrate its necessity for L^q-subellipticity, but will show that it is a reasonable assumption by demonstrating its necessity for certain notions of subellipticity related to Dirichlet and Neumann problems for the balls $B(x, r)$. Indeed, the Sobolev inequality in the third condition in Theorem 10 is almost necessary for the following variant of the notion of L^q-subellipticity. We say that $Q(x, \xi)$ is L^q-subelliptic *relative to the homogeneous Dirichlet problem for the balls $B(x, r)$*, if we assume existence of weak solutions to the homogeneous Dirichlet problem for the balls $B = B(x, r)$,

$$\begin{cases} Lu = f & in \ B \\ u = 0 & on \ \partial B \end{cases},$$

where $L = \nabla'Q(x)\nabla$ is the operator with symbol $Q(x, \xi) = \xi'Q(x)\xi$ and $f \in L^{\frac{q}{2}}(B)$ (in analogy with the elliptic case as treated in Chapter 8 of [**14**]), and if we also assume the following global boundedness estimate for these weak solutions to the above Dirichlet problem,

$$\sup_{z \in B} |u(z)| \le C \left(\int_B |f|^{\frac{q}{2}} \right)^{\frac{2}{q}}$$

(in analogy with Theorem 8.16 of [14] in the elliptic case). We are being deliberately vague regarding the precise definition of these weak solutions, since we have not yet obtained positive results in this direction (we conjecture that there *is* a definition of weak solution such that $\mathcal{Q}(x,\xi)$ is L^q-subelliptic relative to the homogeneous Dirichlet problem for the balls $B(x,r)$ under the hypotheses of Theorem 10 - see subsection 6.7 of the appendix). However, these and related properties have been obtained in some cases in Gutiérrez and Lanconelli [19] - see Theorem 3.1 and Proposition 2.4 there.[1] We prove in Lemma 102 of the appendix that the Sobolev inequality (15) with $\sigma = \frac{q}{q-2}$ is necessary for the L^q-subellipticity of $\mathcal{Q}(x,\xi)$ relative to the homogeneous Dirichlet problem for the balls $B(x,r)$ (in the classical $W^{1,2}$ weak sense), provided $2 < q < Q_*$, where Q_* denotes the lower dimension of the balls $B(x,r)$ ($Q_* \geq n$ if $B(x,r) \subset D(x,Cr)$). Note that the Sobolev inequality (15), together with the other hypotheses of Theorem 10, implies that $\mathcal{Q}(x,\xi)$ is L^q-subelliptic for $q > \max\{Q^*, 2\sigma'\}$. Thus in the special case where the upper and lower dimensions Q^* and Q_* coincide with $2\sigma'$ (this is the case for the homogeneous groups in section 5 of Chapter XIII of Stein [42]), the Sobolev inequality (15) with $\sigma = \frac{Q_*-\varepsilon}{Q_*-\varepsilon-2}$ and $B(x,r) = K(x,r)$ is necessary for $L^{Q_*-\varepsilon}$-subellipticity of the form $\mathcal{Q}$, and the Sobolev inequality (15) with $\sigma = \frac{Q_*+\varepsilon}{Q_*+\varepsilon-2}$ and $B(x,r) = K(x,r)$ is sufficient for the $L^{Q_*+\varepsilon}$-subellipticity of $\mathcal{Q}$ for $\varepsilon > 0$ arbitrarily small, provided conditions 1 and 2 of Theorem 10 are in force. The questions of global boundedness, and existence of weak solutions to degenerate Dirichlet problems, whose study has been initiated in [19], will be taken up in a subsequent paper.

The Poincaré inequality in the third condition in Theorem 10 is almost necessary for a variant of the notion of hypoellipticity. We say that $\mathcal{Q}(x,\xi)$ is L^2-hypoelliptic *relative to the homogeneous Neumann problem for the balls* $B(x,r)$, if we assume existence of weak solutions to the homogeneous Neumann problem for the balls $B = B(x,r)$,

$$(29) \qquad \begin{cases} Lu &= f \quad in\ B \\ \mathbf{n}_Q u &= 0 \quad on\ \partial B \end{cases},$$

where $L = \nabla'Q(x)\nabla$ is the operator with symbol $\mathcal{Q}(x,\xi) = \xi'Q(x)\xi$, the boundary differential operator is $\mathbf{n}_Q = \mathbf{n}'Q(x)\nabla$ where $\mathbf{n}$ is the unit outward normal to ∂B, and $f \in W^{1,2}(B)$ satisfies the compatibility condition $\int_B f = 0$; and if we also assume the following natural hypoelliptic estimate for these weak solutions to the above Neumann problem:

$$(30) \qquad \|u\|_{L^2(B)} \leq Cr^2 \|f\|_{L^2(B)}.$$

Inequality (30) postulates zero gain in smoothness for u from f, which accounts for the terminology hypoelliptic. Again, we are being deliberately vague regarding the precies definition of weak solution to (29). However, we remind the reader that the classical meaning of weak $W^{1,2}$ solution to the boundary value problem (29) is

$$(31) \qquad -\int_B (\nabla v)' Q \nabla u = \int_B vf, \quad for\ all\ v \in W^{1,2}(B).$$

---

[1] One of our students, Scott Rodney, has obtained positive results in these directions as well.

Note that the test functions v in (31) do not necesssarily vanish at the boundary of B. This weak definition is of course motivated by the divergence theorem

$$(32) \qquad \int_{\partial B} v \mathbf{n}_Q u \;=\; \int_{\partial B} \left[vQ\left(x\right) \nabla u \right] \cdot \mathbf{n} = \int_B div\left[vQ\left(x\right) \nabla u \right]$$
$$= \int_B \left(\nabla v\right)' Q\left(x\right) \nabla u + \int_B vLu,$$

valid for $u \in C^2\left(\overline{B}\right)$ and $v \in C^1\left(\overline{B}\right)$ (see e.g. (8.96) in [**14**] for a discussion of mixed boundary value problems). Note that the compatibility condition $\int_B f = 0$ follows from (31) with $v \equiv 1$. We prove in Lemma 103 of the appendix that the Poincaré inequality (17) is necessary for L^2-hypoellipticity of $\mathcal{Q}$ relative to the homogeneous Neumann problem for the balls $B\left(x,r\right)$ (in the classical $W^{1,2}$ weak sense). The existence of weak solutions to degenerate Neumann problems, and the corresponding estimates, will be taken up in a subsequent paper.

2. Extensions of Hörmander's theorem to rough operators

Now we consider extending Hörmander's commutation theorem to rough vector fields in the diagonal case. Suppose that the operator $L = \nabla' B\left(x\right) \nabla$ satisfies (27), i.e.

$$(33) \qquad c_{sym} \sum_{j=1}^{n} \left(a_j\left(x\right) \xi_j\right)^2 \leq \xi' B\left(x\right) \xi \leq C_{sym} \sum_{j=1}^{n} \left(a_j\left(x\right) \xi_j\right)^2, \qquad x \in \Omega, \xi \in \mathbb{R}^n,$$

where $a_j\left(x\right)$ is nonnegative and continuous on Ω. We seek conditions on the weights $a_j\left(x\right)$ in order that any operator $L = \nabla' B\left(x\right) \nabla$ satisfying (33) is subelliptic in Ω according to Definition 6.

DEFINITION 11. *Let $q \in [2, \infty]$. A collection $\mathcal{X} = \{X_j\}_{j=1}^{m}$ of vector fields is L^q-subelliptic in Ω if the quadratic form $\mathcal{Q}\left(x, \xi\right) = \sum_{j=1}^{m} \left(X_j\left(x\right) \cdot \xi\right)^2$ (the principal symbol of $\sum_{j=1}^{m} X_j' X_j$) is L^q-subelliptic in Ω according to Definition 7. We say that the collection $\mathcal{X} = \{X_j\}_{j=1}^{m}$ of vector fields is subelliptic in Ω if it is L^∞-subelliptic in Ω.*

Subellipticity problems for rough vector fields have been considered previously in Franchi [**8**] and Franchi and Lanconelli [**9**], as well as elsewhere (see [**19**] and [**12**] and the references given there). The theorems obtained in [**8**] and [**9**] involve conditions on integral curves or apply to special choices of $a_j\left(x\right)$, except in two dimensions where some of our results essentially coincide with those in [**8**]. See below for a more detailed discussion.

2.1. Diagonal Lipschitz vector fields. Using Theorem 8, we can obtain a general subellipticity theorem for a collection of diagonal Lipschitz vector fields $\mathcal{X} = \{X_j\}_{j=1}^{m}$ that are adapted to a homogeneous space $\mathcal{B}$ of balls (Definition 32) in the sense that the coefficients of the vector fields are sufficiently large compared to the corresponding side lengths of the balls. We now introduce some specialized terminology necessary to state a version of this result. A precise and sharper statement is given in Theorem 82 below when the definition of a prehomogeneous space (Definition 34) is available.

Let $\mathcal{B} = \{B(x,r) : x \in \Omega, 0 < r < \infty\}$ be a homogeneous space on Ω and let $\widetilde{\mathcal{B}}$ be the collection of smallest closed rectangles

$$\widetilde{B}(x,r) = \prod_{k=1}^{n} [x_k - B_k(x,r), x_k + B_k(x,r)]$$

containing $B(x,r)$. We say that $\mathcal{B}$ is Ω-locally equivalent to $\widetilde{\mathcal{B}}$ if there are constants $\delta > 0$ and $C \geq 1$ such that

$$B(x,r) \subset \widetilde{B}(x,r) \subset B(x,Cr), \qquad B(x,r) \in \mathcal{B} \text{ with } x \in \Omega, \, 0 < Cr < \delta \, dist(x, \partial\Omega).$$

More generally, we say two families $\mathcal{B}_j = \{B_j(x,r) : x \in \Omega, 0 < r < \infty\}$ for $j = 1,2$ are Ω-locally equivalent if there are constants $\delta > 0$ and $C \geq 1$ such that

$$\begin{aligned} B_1(x,r) &\subset B_2(x,Cr) \text{ and } B_2(x,r) \subset B_1(x,Cr) \\ \text{for } x &\in \Omega, \, 0 < Cr < \delta \, dist(x, \partial\Omega). \end{aligned}$$

Given a collection $\mathcal{X} = \{X_j\}_{j=1}^{m}$ of vector fields on Ω,

(34) $$X_1 = \frac{\partial}{\partial x_1}, X_2 = a_2(x)\frac{\partial}{\partial x_2}, ..., X_n = a_n(x)\frac{\partial}{\partial x_n},$$

let

$$A_j(x,t) = \int_0^t |a_j(x_1 + s, x_2, \ldots, x_n)| \, ds, \qquad 1 \leq j \leq n,$$

provided the segment joining x and $x + (t, 0, ..., 0)$ lies in Ω. We say the vector fields $\mathcal{X} = \{X_j\}_{j=1}^{m}$ are *adapted* to the homogeneous space $\mathcal{B}$ if

$$C^{-1} \leq \frac{B_1(x,r)}{r} \leq C, \qquad B(x,r) \in \mathcal{B} \text{ with } x \in \Omega, \, 0 < r < \delta \, dist(x, \partial\Omega),$$

and if, for every $0 < \alpha < 1$, there is a positive constant ε such that for all balls $B(x,r) \in \mathcal{B}$ with $x \in \Omega$, $0 < r < \delta \, dist(x, \partial\Omega)$, there is a subset Ω of $\widetilde{B}(x,r)$ with $|\Omega| \geq \alpha \left| \widetilde{B}(x,r) \right|$ and satisfying

$$A_j(z,r) \geq \varepsilon B_j(x,r), \qquad \text{for } z \in \Omega, 1 \leq j \leq n.$$

Since the operators

$$\sum_{j=1}^{n} (a_j(x)\frac{\partial}{\partial x_j})^2 \quad \text{and} \quad \sum_{j=1}^{n} (|a_j(x)|\frac{\partial}{\partial x_j})^2$$

have the same principal symbol, we may restrict consideration to the case when a_j is nonnegative in (34).

THEOREM 12. *Assume that a_i is nonnegative and continuous for x in Ω, doubling in x_1 uniformly in $x_2, \ldots, x_n$ and Lipschitz continuous in $x_2, \ldots, x_n$ uniformly in x_1. Suppose that the collection of vector fields $\mathcal{X} = \{X_j(x)\}_{j=1}^{n}$ in (34) is adapted to a homogeneous space $\mathcal{B}$ on Ω that is Ω-locally equivalent to the family $\widetilde{\mathcal{B}}$. Let D be a doubling exponent for the family $\mathcal{B}$. We either assume that the "accumulating sequence of Lipschitz cutoff functions" condition (20) holds for some $p > \max\{D, 4\}$, or we assume that $\mathcal{B}$ is Ω-locally equivalent to the family of subunit balls $\mathcal{K}$. Then $\mathcal{X}$ is L^q-subelliptic in Ω for all $q > D$.*

The flag condition. We now turn to two special and explicit cases of interest in which we will focus on obtaining conditions of commutation type that guarantee Harnack inequalities and Hölder continuity of weak solutions. Our approach will be to construct homogeneous spaces on which our vector fields are adapted, and then to use Theorems 8, 10 and 12. Recall that for smooth vector fields satisfying Hörmander's commutation condition, the control metric was shown to yield a homogeneous space in [**33**]. Our substitute for Hörmander's commutation condition in Definition 1 in the case of *rough diagonal* vector fields $X_j = a_j(x)\frac{\partial}{\partial x_j}$, $1 \le j \le n$, is the following *flag condition*.

DEFINITION 13. *A collection of continuous vector fields $X_j = a_j(x)\frac{\partial}{\partial x_j}$, $1 \le j \le n$, satisfies the* flag condition *at $x \in \Omega$ if for each index set $\mathcal{I}$ with $\phi \subset \mathcal{I} \subsetneq \{1,2,...,n\}$, there is $j \notin \mathcal{I}$ such that for any neighbourhood $\mathcal{N}$ of x in Ω, a_j does not vanish identically on $(x + \mathcal{V}_\mathcal{I}) \cap \mathcal{N}$ where $\mathcal{V}_\phi = \{0\}$ and $\mathcal{V}_\mathcal{I} = span\{\mathbf{e}_i : i \in \mathcal{I}\}$, $\mathbf{e}_i = (0,...,0,1,0...,0)$ with 1 in the ith position. The vector fields X_j satisfy the flag condition in Ω if they satisfy the flag condition at every point $x \in \Omega$.*

REMARK 14. *An equivalent formulation of the flag condition at x is: there is an increasing sequence of vector spaces*

$$\{0\} = \mathcal{V}_0 \subsetneq \mathcal{V}_1 \subsetneq ...\mathcal{V}_j \subsetneq \mathcal{V}_{j+1} \subsetneq ...\mathcal{V}_m = \mathbb{R}^n,$$

and an increasing sequence of index sets

$$\phi \ne \mathcal{I}_1 \subsetneq ...\mathcal{I}_j \subsetneq \mathcal{I}_{j+1} \subsetneq ...\mathcal{I}_m = \{1,2,...,n\},$$

such that $\mathcal{V}_j = span\{\mathbf{e}_i : i \in \mathcal{I}_j\}$ for $1 \le j \le n$, and a_i does not vanish identically on $(x + \mathcal{V}_j) \cap \mathcal{N}$ for any neighbourhood $\mathcal{N}$ of x in Ω if $i \in \mathcal{I}_{j+1}$, $j \ge 0$.

An increasing sequence of vector spaces $\{\mathcal{V}_j\}_{j=1}^n$ as in Remark 14 is a *flag* at x (see [**32**] for this terminology) and is a *minimal flag* if the length m is a minimum. Remark 30 in section 2.1 shows that a minimal flag is unique. The equivalence mentioned in Remark 14 is easy. Indeed, assume first that for each index set $\mathcal{I}$ with $\phi \subset \mathcal{I} \subsetneq \{1,2,...,n\}$, there is $j \notin \mathcal{I}$ such that a_j does not vanish identically on $(x + \mathcal{V}_\mathcal{I}) \cap \mathcal{N}$ for any neighbourhood $\mathcal{N}$ of x in Ω. Then a flag of length $m = n$ is easily constructed by induction. Conversely, suppose we are given a flag of vector spaces $\{\mathcal{V}_j\}_{j=1}^m$ at $x \in \Omega$. Let $\phi \subset \mathcal{I} \subsetneq \{1,2,...,n\}$, and suppose in order to derive a contradiction, that $a_j \equiv 0$ on $(x + \mathcal{V}_\mathcal{I}) \cap \mathcal{N}$ for some neighbourhood $\mathcal{N}$ of x in Ω, for each $j \notin \mathcal{I}$. In particular, $a_j \equiv 0$ on $(x + \mathcal{V}_0) \cap \mathcal{N}$, i.e. $a_j(x) = 0$, for $j \notin \mathcal{I}$, and it follows that $\mathcal{V}_1 \subset \mathcal{V}_\mathcal{I}$ since $a_j(x) \ne 0$ if $j \in \mathcal{I}_1$. Thus $a_j \equiv 0$ on $(x + \mathcal{V}_1) \cap \mathcal{N}$ for $j \notin \mathcal{I}$, and it follows that $\mathcal{V}_2 \subset \mathcal{V}_\mathcal{I}$. By induction we obtain $\mathbb{R}^n = \mathcal{V}_m \subset \mathcal{V}_\mathcal{I}$, the required contradiction since $\mathcal{I}$ is a proper subset of $\{1,2,...,n\}$.

We now show that if the vector fields $X_j = a_j(x)\frac{\partial}{\partial x_j}$ are Lipschitz continuous, then the flag condition is necessary for subellipticity.

PROPOSITION 15. *Suppose the vector fields $X_j = a_j(x)\frac{\partial}{\partial x_j}$, $1 \le j \le n$, are Lipschitz and that $\mathcal{X} = \{X_j\}_{j=1}^n$ is subelliptic in Ω. Then the flag condition in Definition 13 holds in Ω.*

PROOF. If the flag condition fails at $x = 0$ for $\mathcal{I} = \{m+1, m+2, ..., n\}$ with some $1 \le m < n$, then there exists a neighbourhood $\mathcal{N}$ of 0 such that $a_j \equiv 0$ on

$$\widetilde{\mathcal{V}_\mathcal{I}} = \{x : x_i = 0, 1 \le i \le m\} \cap \Omega \cap \mathcal{N}, \quad 1 \le j \le m.$$

Define φ on $[0, \infty)$ by

$$\varphi(t) = \begin{cases} t, & 0 \leq t \leq 1 \\ 1, & t > 1 \end{cases}.$$

For $\varepsilon > 0$ sufficiently small, set $u^\varepsilon(x) = \varphi\left(\frac{|x'|}{\varepsilon}\right)$ where $x' = (x_1, ..., x_m)$, and define $f_j^\varepsilon = X_j u^\varepsilon = a_j \frac{\partial u^\varepsilon}{\partial x_j}$. Since a_j is Lipschitz and vanishes on $\widetilde{\mathcal{V}_\mathcal{I}}$ for $1 \leq j \leq m$, we have $|a_j(x)| \leq C |x'|$ for $1 \leq j \leq m$, and so also

$$\left| f_j^\varepsilon(x) \right| = \left| a_j(x) \varphi'\left(\frac{|x'|}{\varepsilon}\right) \frac{1}{\varepsilon} \frac{\partial |x'|}{\partial x_j} \right| \leq C, \qquad 1 \leq j \leq n.$$

Thus u^ε is a $W^{1,2}(\Omega)$ weak solution of

$$L u^\varepsilon = \sum_{j=1}^n T_j' f_j^\varepsilon,$$

where $L = \sum_{j=1}^n X_j' X_j$ satisfies (33), $T_j = X_j$ is subunit with respect to the matrix

$$\begin{pmatrix} a_1(x)^2 & \cdots & 0 \\ \vdots & \ddots & \vdots \\ 0 & \cdots & a_n(x)^2 \end{pmatrix}$$

corresponding to L, and $\|u^\varepsilon\|_2$, $\|f_j^\varepsilon\|_\infty$ are uniformly bounded for $0 < \varepsilon < 1$. Nevertheless, the u^ε are not uniformly bounded in any Hölder space, and this shows that $\mathcal{X} = \{X_j\}_{j=1}^n$ is not subelliptic in Ω.

In the next section, we show that in the case the $a_j(x)$ are real-analytic, then the vector fields $\{X_j\}_{j=1}^n$ satisfy the flag condition in Definition 13 if and only if they satisfy Hörmander's commutation condition in Definition 1. In order to state the first of our two generalizations of Hörmander's commutation theorem, we recall the definition of the *reverse Hölder condition*, referred to as RH_∞ in [8] and elsewhere.

DEFINITION 16. *A nonnegative function $a(t)$ defined on an open subset J of $\mathbb{R}$ satisfies the* reverse Hölder condition of infinite order *if*

$$\operatorname*{ess\,sup}_{t \in I} a(t) \leq C \frac{1}{|I|} \int_I a(t)\, dt$$

for all intervals $I \subset J$.

THEOREM 17. *Suppose for $1 \leq j \leq n$ that $a_j(x)$ is nonnegative and Lipschitz continuous on a domain $\Omega \subset \mathbb{R}^n$, and reverse Hölder in each variable x_i with $i \neq j$, uniformly in the remaining variables. Then the set $\mathcal{X}$ of vector fields $X_j = a_j \frac{\partial}{\partial x_j}$, $1 \leq j \leq n$, is subelliptic in Ω if and only if $\mathcal{X}$ satisfies the flag condition in Definition 13 in Ω. In the case that the flag condition in Definition 13 holds in Ω, there is $Q \in [n, \infty)$ depending only on the Lipschitz and reverse Hölder constants of the a_j such that $\mathcal{X}$ is L^q-subelliptic in Ω for all $q > Q$.*

The constant Q can be taken to be a doubling exponent D of the flag balls $B(x, r)$ defined in subsection 2.2.2 below; see also 5.2.1. A doubling exponent D is given by the inequality

$$(35) \qquad |B(x,r)| \leq C\left(\frac{r}{t}\right)^{D} |B(y,t)|, \qquad \text{whenever } B(x,r) \supset B(y,t)$$
$$0 < r < \delta \, dist(x, \partial\Omega), 0 < t < \delta \, dist(y, \partial\Omega).$$

See Lemma 73 in section 4 for details.

The noninterference condition. In order to motivate our second generalization of Hörmander's theorem, we observe that an operator $L = \nabla' B(x) \nabla$ satisfying the quadratic form condition in Definition (33) with respect to continuous vector fields $X_j = a_j(x)\frac{\partial}{\partial x_j}$ that satisfy the flag condition in Definition 13 is elliptic if and only if m can be taken to be 1 in Definition 13 for all $x \in \Omega$. The simplest nonelliptic case thus occurs when for every $x \in \Omega$, a minimal flag has length m at most 2, and when the length m is exactly 2, the index set $\mathcal{I}_1$ can be chosen to be a singleton, i.e. $\#\mathcal{I}_1 = 1$. Of course there must be at least one point x where $m = 2$, since otherwise L is elliptic. For example, in the case that $m = 2$ and $\mathcal{I}_1 = \{1\}$ at a point y, we assume that in some neighbourhood of y we have

$$(36) \qquad a_1(x) \geq c > 0,$$
$$a_j(\cdot, x_2, ..., x_n) > 0 \text{ on a dense subset for all } x_2, ..., x_n \text{ and } 2 \leq j \leq n.$$

We will also assume that a_j is nonnegative and reverse Hölder of infinite order in x_1 for $j \geq 2$ (but *not* in x_i for $i \neq 1$). Moreover, we will suppose there is a positive constant $C_{\max}$ such that

$$(37) \qquad \max_{1 \leq j \leq n} \sup_{x \in \Omega} a_j(x) \leq C_{\max}.$$

In this case there is a substitute for the requirement that a_j satisfies the reverse Hölder condition in *all* of the variables x_i with $i \neq j$, in order to obtain L^q-subellipticity for L.

To state this condition, we first define open rectangular boxes of radius r and center x by

$$(38) \qquad A(x,r) = \prod_{j=1}^{n} (x_j - A_j(x,r), x_j + A_j(x,r)),$$

where half the jth side length of $A(x,r)$ is given by

$$(39) \qquad A_j(x,r) = \int_0^r a_j(x_1 + t, x_2, ..., x_n)\, dt,$$

provided the segment joining x and $x + (r, 0, ..., 0))$ lies in Ω. Note that $A_j(x,r) > 0$ for $j \geq 2$, $r > 0$ by (36), and thus we may rewrite (36) as the following nondegeneracy condition,

$$(40) \qquad A_1(x,r) \geq cr \text{ and } A_j(x,r) > 0 \text{ for } x \in \Omega, 0 < r < \delta \, dist(x, \partial\Omega),$$

where δ is chosen small enough that $A_j(x,r)$ is defined for $x \in \Omega$, $0 < r < \delta \, dist(x, \partial\Omega)$. To simplify notation, we will always assume from now on that the definitions of the $A_j(x,r)$ are with respect to the variable x_1, and that (40) holds. The reader can easily supply the more general statements and proofs. Even when the a_j are reverse Hölder of infinite order in the x_1 variable uniformly in $x_2, ..., x_n$, the rectangular boxes $A(x,r)$ do not in general form a homogeneous space structure as in subsection 2.2, since the engulfing property can fail as evidenced by the three dimensional example

$$
\begin{aligned}
a_1(x) &= 1, \\
a_2(x) &= 3x_1^2, \\
a_3(x) &= 9x_1^8 + x_2^2.
\end{aligned}
$$

See Example 49 in subsection 2.3.1 for details. This problem can be avoided if the following *noninterference condition* is assumed, which has the consequence of forcing the rectangles $A(x,r)$ to form a (local) homogeneous space structure in the sense of Definition 32 below, when the a_j are doubling weights in the x_1 variable.

DEFINITION 18. *Suppose the continuous vector fields* $X_j = a_j \frac{\partial}{\partial x_j}$ *satisfy* (40) *in* Ω. *We say that the vector fields* $\{X_j\}_{j=1}^n$ *satisfy a* noninterference condition *in* Ω *if there are positive constants* C_c *and* δ *such that*

$$
C_c^{-1} A_j(x,r) \le A_j(z,r) \le C_c A_j(x,r), \quad z \in A(x,r), x \in \Omega, 0 < r < \delta dist(x, \partial\Omega),
$$

for $1 \le j \le n$.

We remark that if δ is sufficiently small depending on $C_{\max}$ in (37), then $A_j(z,r)$ is defined for $z \in A(x,r)$ when $0 < r < \delta\, dist(x, \partial\Omega)$. Since the above inequality is automatic for $j = 1$, we may restrict j to the range $2 \le j \le n$. In particular, as we show in Lemma 109 of the appendix, the noninterference condition in Definition 18 is implied by the following *strong noninterference condition* for Lipschitz continuous vector fields in Ω relative to the boxes $A(x,r)$:
(41)

$$
r \left\{ \sup_{z \in A(x,r)} \left| \frac{\partial a_j}{\partial x_i}(z) \right| \right\} A_i(x,r) \le C A_j(x,r), \quad x \in \Omega, 0 < r < \delta\, dist(x, \partial\Omega),
$$

for $2 \le i, j \le n$.

REMARK 19. *A comment regarding the supremum in* (41) *is in order. If* $a_j(z)$ *is Lipschitz continuous in* z *with constant* C, *then* $a_j(z_1, ..., z_n)$ *is Lipschitz continuous in each variable* z_i *separately. Thus for each fixed* $z_1, ..., z_{i-1}, z_{i+1}, ..., z_n$ *the partial derivative* $\frac{\partial a_j}{\partial z_i}(z_1, ..., z_n)$ *exists for a.e.* z_i *and has absolute value bounded by* C *whenever it exists. We will always interpret the expression* $\sup_{z \in A(x,r)} \left| \frac{\partial a_j}{\partial z_i}(z) \right|$ *to mean the supremum of the Lipschitz constants of the functions* $z_i \to a_j(z_1, ..., z_n)$ *for* $(z_1, ..., z_n)$ *such that* $(z_1, ..., z_n) \in A(x,r)$.

Note that the strong noninterference condition (41) is automatic if $i = j$ or for any i, j with $A_i(x,r) \approx A_j(x,r)$, and thus vacuously true in dimension $n = 2$. Under some natural hypotheses, the same is actually true of Definition 18, and this is proved in Lemma 111 of the appendix.

A simple example satisfying Definition 18 is $a_1 \equiv 1$ and $a_j(x) = |x|^{p_j}$ for $2 \le j \le n$, where p_j is a real number greater than or equal to one. In this case, $A_j(x,r) \approx r(r + |x|)^{p_j}$.

THEOREM 20. *Suppose for* $1 \le j \le n$ *that* $a_j(x)$ *is nonnegative and Lipschitz continuous on a domain* $\Omega \subset \mathbb{R}^n$, *and that the vector fields* $X_j = a_j \frac{\partial}{\partial x_j}$ *satisfy* (40) *in* Ω. *We also assume the noninterference condition in Definition* 18. *Suppose moreover that each* a_j *with* $2 \le j \le n$ *is reverse Hölder of infinite order in* *the variable x_1, uniformly in the remaining variables. Then there is $Q \in [n, \infty)$ depending only on the Lipschitz and reverse Hölder constants of the a_j such that the set $\mathcal{X} = \{X_j\}_{j=1}^n$ of vector fields is L^q-subelliptic in Ω for all $q > Q$.*

As before, the constant Q can be taken to be a doubling exponent D of the rectangles $A(x, r)$. Again, see (45) below and Lemma 73 in section 3. We will often refer to the open rectangles $A(x, r)$ defined in (38) as *noninterference balls*.

In particular, the following special case arises in connection with the Monge-Ampère equation discussed in subsection 1.3 on applications below. If $a_1(x) \equiv 1$ and $a_i(x) = a_j(x)$ for $2 \le i, j \le n$ in Theorem 20, we have the following corollary.

COROLLARY 21. *Suppose that $a(x)$ is nonnegative and Lipschitz continuous on a domain $\Omega \subset \mathbb{R}^n$, and reverse Hölder of infinite order in the variable x_1 uniformly in $x_2, ..., x_n$. Suppose that $a(x_1, x_2, ..., x_n)$ doesn't vanish identically in x_1 for any $x_2, ..., x_n$. Then there is $Q \in [n, \infty)$ depending only on the Lipschitz and reverse Hölder constants of the a_j such that the quadratic form*

$$(42) \qquad \mathcal{Q}(x, \xi) = \xi_1^2 + a(x)^2 \left(\xi_2^2 + ... + \xi_n^2 \right)$$

is L^q-subelliptic in Ω for all $q > Q$.

As noted above, the constant Q can be taken to be a doubling exponent D of the rectangles $A(x, r)$ as in (45). Corollary 21 is a corollary of Theorem 20 since by Lemma 111 in subsection 6.6 of the appendix, the noninterference condition in Definition 18 holds automatically if $a_i = a_j$ for $2 \le i, j \le n$. A stonger result is given in Corollary 26 in the next subsection.

Franchi [8] and Franchi and Lanconelli [9] have also obtained versions of Theorem 17 and 20 and Corollary 21. In [9], the a_j considered are products of special one-dimensional functions that behave like monomials, and in [8] a strong form of the reverse Hölder condition is assumed along integral curves γ_u of the vector fields $T_{\mathbf{u}} = \sum_{j=1}^n u_j a_j(x) \frac{\partial}{\partial x_j}$, $0 < |u_j| < 1$, namely,

$$(43)$$

$$\sup_{z \in C\widetilde{K}(x,r)} a_j(z)$$

$$\le C \inf \left\{ \frac{1}{r} \int_0^r a_j(\gamma_u(t)) \, dt : \gamma_u(t) = T_{\mathbf{u}}(\gamma_u(t)), |u_j| \ge \varepsilon > 0, \gamma_u(0) = x \right\},$$

where $K(x, r)$ are the balls corresponding to the subunit metric, and $\widetilde{K}(x, r)$ denotes the smallest closed rectangular box with edges parallel to the coordinate axes that contains $K(x, r)$. In $n = 3$ dimensions, if we take $0 < \beta_1 < ... < \beta_N$ and set

$$
\begin{aligned}
a_1(x) &= 1, \\
a_2(x) &= 3x_1^2, \\
a_3(x) &= \prod_{j=1}^N \left(x_2 - \beta_j x_1^3 \right)^2,
\end{aligned}
$$

then the hypotheses of Theorem 17 hold, yet $\int_0^r a_3(\gamma_u(t)) \, dt = 0$ for $\frac{u_2}{u_1} \in \{\beta_1, ..., \beta_N\}$, $\gamma_u(0) = 0$, and so the strong form of the reverse Hölder condition (43) fails. Nevertheless, with some work, one can show that (43) does hold under the special hypotheses of Corollary 21.

2.2. Sharper technical theorems. Due to the local nature of the L^q-subelliptic conclusion in Theorems 17 and 20, we may assume without loss of generality that $a_1(x) \equiv 1$, i.e. $X_1 = \frac{\partial}{\partial x_1}$. Indeed, fix $x_0 \in \Omega$. We may suppose that $a_1(x_0) = \max_{1 \leq j \leq n} a_j(x_0) > 0$, and then define $\widetilde{\Omega} = \{x \in \Omega : a_1(x) > \frac{1}{2}a_1(x_0)\}$ and $\widetilde{a}_j(x) = \frac{a_j(x)}{a_1(x)}$ for $x \in \widetilde{\Omega}$. It is clear that the vector fields $\widetilde{X}_j = \widetilde{a}_j(x)\frac{\partial}{\partial x_j}$, $1 \leq j \leq n$, satisfy in $\widetilde{\Omega}$ the same quadratic form, Lipschitz continuity, reverse Hölder, flag and/or noninterference conditions as the vector fields $X_j = a_j(x)\frac{\partial}{\partial x_j}$, $1 \leq j \leq n$, do in Ω, and with comparable constants depending on x_0. In this way, we have distinguished a direction, namely the x_1 direction, in which curves of unit Euclidean speed are subunit.

Having established this, it is now possible to relax the Lipschitz continuity of the remaining coefficients a_j, $2 \leq j \leq n$, in this distinguished direction, still assuming Lipschitz continuity in the transverse directions (transverse continuity is essential as evidenced by the pathological Example 48 in the next section). Moreover, it is also possible to relax the reverse Hölder continuity of the coefficients a_j in the x_1 variable. For this, we need the weaker reverse Hölder condition of order $p \in (1, \infty)$:

DEFINITION 22. *Let $1 < p < \infty$. A nonnegative function $a(t)$ defined on an open subset J of $\mathbb{R}$, satisfies the reverse Hölder condition of order p, denoted $a \in RH_p$, if*

$$(44) \qquad \left(\frac{1}{|I|}\int_I a(t)^p\, dt\right)^{\frac{1}{p}} \leq C\frac{1}{|I|}\int_I a(t)\, dt,$$

for all intervals $I \subset J$.

This results in the following sharper, but more technical, sufficient conditions for subellipticity.

THEOREM 23. *Suppose that $a_1(x) \equiv 1$ and $a_j(x)$ is continuous and nonnegative on a domain $\Omega \subset \mathbb{R}^n$, $2 \leq j \leq n$. Moreover, suppose a_j is Lipschitz continuous in $x_2, ..., x_n$ uniformly in x_1, reverse Hölder of infinite order in each variable x_i with $i \neq 1, j$, uniformly in the remaining variables, and reverse Holder of order $p > \max\{D, 4\}$ in the variable x_1, uniformly in the remaining variables, where D is a doubling exponent for the flag balls as in (35). If the set $\mathcal{X}$ of vector fields $X_j = a_j\frac{\partial}{\partial x_j}$, $1 \leq j \leq n$, satisfies the flag condition in Definition 13 in Ω, then $\mathcal{X}$ is L^q-subelliptic in Ω for all $q > D$.*

In order to state the sharpened form of Theorem 20, we need a doubling exponent D for the noninterference balls $A(x, r)$:

$$(45) \qquad |A(x, r)| \; \leq \; C\left(\frac{r}{t}\right)^D |A(y, t)|, \qquad \text{whenever } A(x, r) \supset A(y, t),$$
$$0 \; < \; r < \delta\, dist(x, \partial\Omega), 0 < t < \delta\, dist(y, \partial\Omega).$$

See Lemma 73.

THEOREM 24. *Suppose that $a_1(x) \equiv 1$ and $a_j(x)$ is continuous and nonnegative on a domain $\Omega \subset \mathbb{R}^n$, $2 \leq j \leq n$. Let D be a doubling exponent in (45). Suppose that both the nondegeneracy condition (40) and the noninterference condition in Definition 18 hold, and that each a_j satisfies the RH_p condition (44) for some $p > \max\{D, 4\}$ in the variable x_1, uniformly in $x_2, \ldots, x_n$, and that each a_j is*

Lipschitz continuous in $x_2, \ldots, x_n$, *uniformly in* x_1. *Then the set* $\mathcal{X} = \{X_j\}_{j=1}^{n}$ *of vector fields is* L^q-*subelliptic in* Ω *for all* $q > D$.

There are two differences between Theorems 17 and 23, and between Theorems 20 and 24. The a_j are no longer assumed Lipschitz continuous in the distinguished variable x_1, and are no longer assumed reverse Hölder of infinite order in x_1.

EXAMPLE 25. *See subsection 6.4 of the appendix for examples of Lipschitz continuous functions* a_j *that satisfy (44) for* $p > \max\{D, 4\}$, *yet fail to be reverse Hölder of infinite order. By Proposition 104, such examples give rise to a homogeneous space of noninterference balls* $A(x, r)$ *that are* not *equivalent to the subunit balls* $K(x, r)$ *of Fefferman and Phong* [7] *and Nagel, Stein and Wainger* [33]. *Nevertheless, by Theorem 24, the corresponding set of vector fields* $\left\{ a_j \frac{\partial}{\partial x_j} \right\}_{j=1}^{n}$ *is still* L^q-*subelliptic for large* q.

We close with a strengthening of Corollary 21.

COROLLARY 26. *Suppose that* $a(x)$ *is continuous and nonnegative on a domain* $\Omega \subset \mathbb{R}^n$, *Lipschitz continuous in the variables* $x_2, \ldots x_n$ *uniformly in* x_1, *and reverse Hölder of order* $p > \max\{D, 4\}$ *in the variable* x_1 *uniformly in* $x_2, \ldots, x_n$, *where* $D = 1 + (n - 1)d$ *and* d *is the doubling order in*
(46)
$$\int_J a(x)\, dx_1 \leq C \left(\frac{|J|}{|I|} \right)^d \int_I a(x)\, dx_1, \quad \text{whenever } I \subset J, \text{ uniformly in } x_2, \ldots, x_n,$$
so that D *is as in (45). Suppose that* $a(\cdot, x_2, \ldots, x_n)$ *doesn't vanish identically in any nontrivial interval for any* $x_2, \ldots, x_n$. *Then the quadratic form*
$$\mathcal{Q}(x, \xi) = \xi_1^2 + a(x)^2 \left(\xi_2^2 + \ldots + \xi_n^2 \right)$$
is L^q-*subelliptic in* Ω *for all* $q > D$.

This is a corollary of Theorem 24 since the noninterference condition in Definition 18 is automatic by Lemma 111 if the a_j coincide for $2 \leq j \leq n$.

2.3. Connections between the theorems. Here we discuss the connection of Theorems 12, 17 and 20 with Theorems 8 and 10. Following the approach pioneered in [33], our strategy for proving Theorems 17 and 20 is to use the vector fields $X_j = a_j(x) \frac{\partial}{\partial x_j}$, $1 \leq j \leq n$, to construct balls $B(x, r)$ and $A(x, r)$ that satisfy a δ-local prehomogeneous and homogeneous space structure respectively in Ω, relative to Lebesgue measure. See subsection 2.2 for a detailed discussion of prehomogeneous and homogeneous spaces. A key feature of these spaces is the doubling condition
$$
\begin{aligned}
|B(x, 2r)| &\leq C|B(x, r)|, & 0 < r < \infty, \\
|A(x, 2r)| &\leq C|A(x, r)|, & 0 < r < \infty,
\end{aligned}
$$
first established for smooth vector fields satisfying Hörmander's commutation condition in [33]. We also establish a subrepresentation inequality for Lipschitz functions f in terms of their a-gradients $\nabla_a f = \left(a_j \frac{\partial f}{\partial x_j} \right)_{j=1}^{n}$ of the form:
$$
(47) \qquad \left| f(x) - \frac{1}{|B(y, r)|} \int_{B(y, r)} f \right| \leq C \int_{B(y, r)} |\nabla_a f(z)| \frac{d(x, z)}{|B(x, d(x, z))|} dz,
$$

for $x \in B(y, r)$, where $d(x, y)$ is the quasimetric associated with the balls $B(x, r)$, along with a similar result for the balls $A(x, r)$. The main tool used for this is the notion of adaptability of balls to vector fields as used in the statement of Theorem 12. From these facts and others follow Poincaré and Sobolev inequalities, such as those in (15) and (17), as well as the local equivalence of these families of balls with the subunit balls $K(x, r)$ of Fefferman and Phong. At this point, we are able to apply Theorem 10 to obtain subellipticity of the vector fields $\{X_j\}_{j=1}^n$ in Theorems 17 and 20, and to apply Theorem 8 to obtain subellipticity in Theorem 12. These methods yield not only Hölder continuity for weak solutions u to the equation

$$Lu + \mathbf{H}\mathbf{R}u + \mathbf{S}'\mathbf{G}u + Fu = f + \mathbf{T}'\mathbf{g},$$

but also Harnack inequalities for nonnegative weak solutions u to this equation in a ball $B(y, r)$: there exist positive constants $c < 1 < C$ such that

$$\operatorname{ess} \sup_{x \in B(y, cr)} u(x) \leq C \operatorname{ess} \inf_{x \in B(y, cr)} u(x) + r^{2\delta} \|f\|_{\frac{q}{2}} + r^{\delta} \|\mathbf{g}\|_q,$$

where $q(1 - \delta) > D$ and D is the doubling exponent for the balls in question. See Theorem 62 in section 3 below.

As mentioned above, the balls $B(x, r)$ and $A(x, r)$ both turn out to be locally equivalent to the subunit balls $K(x, r)$ when the continuous vector field coefficients $a_j(x)$ are reverse Hölder of infinite order in the x_1-variable, despite the different constructions used in each case. In fact, assuming only that the continuous $a_j(x)$ are Lipschitz in $x_2, ..., x_n$ and doubling in x_1, uniformly in the remaining variables, the noninterference balls $A(x, r)$ are locally equivalent to the subunit balls $K(x, r)$ if and only if the $a_j(x)$ are reverse Hölder of infinite order in the x_1-variable. Thus in the event the vector field coefficients $a_j(x)$ are RH_p, but not RH_∞, in the x_1-variable for p sufficiently large, namely p larger than the doubling exponent D of the balls $A(x, r)$, the collection of vector fields $\mathcal{X} = \{X_j\}_{j=1}^n$ is subelliptic despite the nonequivalence of the balls $A(x, r)$ and $K(x, r)$. Moreover, the Fefferman-Phong condition (19) is necessary for the subellipticity of the Lipschitz continuous vector fields $\{X_j\}_{j=1}^n$ in the sense of Definition 11 (and necessary for stable subellipticity if the X_j are merely continuous).

See section 2 and the appendix for proofs of these assertions regarding subunit balls. In section 3, we prove our general result Theorem 8. In section 4 we complete the proof of Theorem 10, extending the theorem of Fefferman and Phong, by establishing the "accumulating sequence of Lipschitz cutoff functions on annuli" condition (20) with $p = \infty$. We next establish a proportional subrepresentation inequality similar to (47), and use this to prove Theorem 12 on vector fields adapted to a homogeneous space structure. Then we use Theorem 12 to reduce the proofs of Theorems 17 and 20, our extensions of Hörmander's theorem in the case of diagonal vector fields, to establishing appropriate homogeneous space structures adapted to $\mathcal{X}$. In section 5, we establish these homogeneous space structures adapted to $\mathcal{X}$ for both the flag balls and noninterference balls. Before proceeding with this plan, we indicate some applications to hypoellipticity of smooth nonlinear equations.

3. Applications to quasilinear equations

In order to apply our rough subelliptic theorems to hypoellipticity of the quasilinear equations (6), and more generally to diagonal systems of this form, we recall the regularity theorems for quasilinear equations in [16] and [34]. As in [16], a symmetric nonnegative Lipschitz matrix $A(y)$ is *subordinate* in a domain $\Omega \subset \mathbb{R}^N$ if

$$(48) \qquad \sum_{j=1}^{n} \left(\sum_{i=1}^{n} \frac{\partial}{\partial y_\ell} a_{ij}(y) \xi_i \right)^2 \leq C\xi' A(y) \xi, \qquad y \in \Omega, \xi \in \mathbb{R}^n, 1 \leq \ell \leq N.$$

THEOREM 27. *(Guan [16]) Suppose that $u \in Lip_1(\Omega)$ is a weak solution of the divergence form equation*

$$\sum_{i,j=1}^{n} \frac{\partial}{\partial x_i} \left(a_{ij}(x, u(x)) \frac{\partial u}{\partial x_j} \right) = f(x), \qquad x \in \Omega,$$

where $a_{ij} \in C^\infty(\Omega \times \mathbb{R})$, $A = [a_{ij}]_{ij=1}^{n}$ is symmetric, nonnegative semidefinite, and subordinate in relatively compact subdomains of $\Omega \times \mathbb{R}$, and $f \in C^\infty(\Omega)$. Let

$$\widetilde{L} = \nabla' \widetilde{A}(x) \nabla = \sum_{i,j=1}^{n} \frac{\partial}{\partial x_i} \widetilde{a_{ij}}(x) \frac{\partial}{\partial x_j}$$

be the linear operator with $\widetilde{a_{ij}}(x) = a_{ij}(x, u(x))$. Suppose that

$$trace [\widetilde{a_{ij}}]_{ij=1}^{n} \geq c > 0$$

in Ω, and that $\widetilde{L}$ is α-subelliptic in Ω for some $\alpha > 0$. Then u is smooth in Ω.

Clearly Theorems 17, 20 and 27 can be combined to yield smoothness of Lipschitz solutions to quasilinear equations of the above type when the linearized operator $\widetilde{L}$ can be shown to satisfy the hypotheses of either Theorem 17 or 20.

We now mention the application of Corollary 21 to the Monge-Ampère equation in [**34**]. We first recall the theorem for quasilinear systems in [**34**]. Note that the unknowns in the system below are vector functions **p** acted on by nonlinear second order operators, and vector functions **v** that are connected to the unknowns **p** via a simple elliptic equation. This flexibility permits application to equations of Monge-Ampère type transformed by a partial Legendre transform. We recall another definition of Guan [**15**]. We say that $L = \nabla' A(x) \nabla$ is *α-elliptic extendible* in Ω for $\alpha > 0$ if for every x_0 and Ω_1 with $x_0 \in \Omega_1 \Subset \Omega$, there is a symmetric smooth nonnegative subordinate matrix $B(x)$ in Ω that vanishes in a neighbourhood $\mathcal{N} \Subset \Omega_1$ of x_0, is elliptic in $\Omega - \Omega_1$, and such that

$$L_\varepsilon = \nabla' (A(x) + B(x) + \varepsilon I) \nabla$$

is α-subelliptic in Ω, uniformly in $0 < \varepsilon < 1$.

The scalar version of the next result with the simpler right-hand side $f(x)$ is due to Guan [**15**], and included in Theorem 27 above.

THEOREM 28. *Suppose that $\mathbf{p} = (p_\ell)_{1 \leq \ell \leq N}, \mathbf{v} = (v_\ell)_{1 \leq \ell \leq N_0} \in Lip_1(\Omega)$ and that $(\mathbf{p}, \mathbf{v})$ is a weak solution of the system*

$$\left\{ \sum_{i,j=1}^{n} \frac{\partial}{\partial x_i} a_{ij}(x, \mathbf{v}, \mathbf{p}) \frac{\partial}{\partial x_j} \right\} p_\ell = h_\ell(x, \mathbf{v}, \mathbf{p}, D\mathbf{p}), \qquad 1 \leq \ell \leq N,$$

$$D\mathbf{v} = \boldsymbol{\Psi}(x, \mathbf{v}, \mathbf{p}),$$

where $a_{ij} \in C^\infty(\Gamma)$, Γ is a subdomain of $\Omega \times \mathbb{R}^{N_0} \times \mathbb{R}^N$, $A(x, \mathbf{v}, \mathbf{p}) = [a_{ij}(x, \mathbf{v}, \mathbf{p})]_{ij=1}^{n}$ is symmetric, nonnegative semidefinite, and subordinate in relatively compact subdomains of Γ, $\mathbf{h} = (h_\ell)_{1 \leq \ell \leq N} \in C^\infty(\Gamma \times \mathbb{R}^{nN})$ and $\boldsymbol{\Psi} \in C^\infty(\Gamma)$. Let $\widetilde{L} =$

$\nabla' \widetilde{A}(x) \nabla = \sum_{i,j=1}^{n} \frac{\partial}{\partial x_i} \widetilde{a_{ij}}(x) \frac{\partial}{\partial x_j}$ *be the scalar linear operator with* $\widetilde{a_{ij}}(x) = a_{ij}(x, \mathbf{v}(x), \mathbf{p}(x))$. *Suppose that* $\widetilde{L}$ *is* α-*elliptic extendible in* Ω *for some* $\alpha > 0$, *that*

$$trace \, [\widetilde{a_{ij}}]_{ij=1}^{n} \geq c > 0 \quad in \; \Omega,$$

and that $\mathbf{h}$ *has the product decomposition*

$$h_\ell(x, \mathbf{v}, \mathbf{p}, D\mathbf{p}) = H_{\ell,0}(x, \mathbf{v}, \mathbf{p}) + \sum_{\mu=1}^{M} H_{\ell,\mu}(x, \mathbf{v}, \mathbf{p}) \, \Phi_{\ell,\mu}(x, \mathbf{v}, \mathbf{p}, D\mathbf{p}), \quad 1 \leq \ell \leq N,$$

with $H_{\ell,\mu}$ *and* $\Phi_{\ell,\mu}$ *smooth functions of their arguments, and where the vector fields*

$$H_{\ell,\mu}(x, \mathbf{v}(x), \mathbf{p}(x)) \frac{\partial}{\partial x_k}$$

are subunit with respect to $\widetilde{A}$ *for* $1 \leq \mu \leq M, 1 \leq \ell \leq N, 1 \leq k \leq n$. *Then* $\mathbf{p}$ *and* $\mathbf{v}$ *are both smooth in* Ω.

Corollary 21 and Theorem 28 apply to the quasilinear system

$$\left\{ \frac{\partial^2}{\partial x_1^2} + \sum_{i,j=2}^{n} \frac{\partial}{\partial x_i} k(x, \mathbf{v}, \mathbf{p}) M_{ij}(\mathbf{p}) \frac{\partial}{\partial x_j} \right\} p_\ell = h_\ell(x, \mathbf{v}, \mathbf{p}, D\mathbf{p}), \quad 1 \leq \ell \leq N,$$

$$D\mathbf{v} = \mathbf{p},$$

that arises from the generalized Monge-Ampère equation,

$$(49) \qquad\qquad \det D^2 u = k(x, u, Du), \quad x \in \Omega,$$

where k is smooth and nonnegative in $\Omega \times \mathbb{R} \times \mathbb{R}^n$, and Ω is a convex domain in $\mathbb{R}^n$. Indeed, using the higher dimensional partial Legendre transform corresponding to a convex $C^{2,1}$ solution u of (49),

$$(50) \qquad\qquad \begin{cases} s &= x_1 \\ t_2 &= \frac{\partial u}{\partial x_2}(x) \\ \vdots & \\ t_n &= \frac{\partial u}{\partial x_n}(x) \end{cases},$$

the vector-valued functions $\mathbf{v} = (v_\ell)_{\ell=2}^{n} = (x_\ell(s, \mathbf{t}))_{\ell=2}^{n}$ and $\mathbf{p} = D\mathbf{v} = \left(\frac{\partial v_i}{\partial t_j}\right)_{\substack{2 \leq i \leq n \\ 1 \leq j \leq n}}$ ($s = t_1$) arising from the inverse transform, satisfy a divergence form quasilinear system

$$(51) \qquad \mathcal{L}\mathbf{p} \equiv \left\{ \frac{\partial^2}{\partial s^2} + \nabla'_t k M(\mathbf{p}) \nabla_t \right\} \mathbf{p} = \mathbf{f}((s, \mathbf{t}), \mathbf{v}, \mathbf{p}, D\mathbf{p}),$$

in the classical weak sense as given in Definition 50 below. Assuming that $u \in C^{2,1}$ and $k(x, u, Du) \approx x_1^{2m} + \psi(x)$ where $\psi(x)^{\frac{1}{2m}}$ is Lipschitz for some $m \in N$, and that $\det [\partial_i \partial_j u]_{i,j=2}^{n} > 0$ so that $M(\mathbf{p})$ is a positive definite matrix, Corollary 21 applies to show that the linearization of (51) is α-subelliptic for some $\alpha > 0$, and then Theorem 28 yields smoothness. We note here that the main feature of Corollary 21 that permits its use with the partial Legendre transform (50) is that no assumption other than Lipschitz is required of $a(x)$ in the variables $x_2, ..., x_n$ which get replaced with the unknown Lipschitz functions $(v_2, ..., v_n)$. Indeed, it is this property that allows us to verify the hypothesis in Theorem 28 that the linear operators $\widetilde{L}$ are *α-elliptic extendible* in Ω for appropriate k. See [**34**] for details.

An easily stated special case is this: If the smooth nonnegative prescribed Gaussian curvature $k_n(x)$ of the graph of a $C^{2,1}$ function $u(x)$ vanishes at a nondegenerate critical point at x_0, then $u(x)$ is smooth near x_0 if (and only if) $k_{n-1}(x_0) > 0$. Here k_j denotes the jth symmetric curvature of u. Again, see [**34**] for details.

We also point out that certain subelliptic quasilinear systems of equations have been considered by Xu and Zuily in [**46**]. They use the Campanato method to treat equations of the form

$$(52) \qquad \left\{ \sum_{i,j=1}^{m} X_i' M_{ij}(x, \mathbf{p}) X_j \right\} p_\ell = f_\ell(x, \mathbf{p}, D\mathbf{p}), \qquad 1 \le \ell \le N,$$

where $\mathbf{p} = (p_1, ..., p_N)$ is assumed continuous and $D\mathbf{p}$ locally square integrable, $M = [M_{ij}(x, \mathbf{p})]_{i,j=1}^{m}$ is smooth and elliptic, $\mathbf{f} = (f_1, ..., f_N)$ is smooth and has at most quadratic growth in $D\mathbf{p}$, and $\{X_j\}_{j=1}^{m}$ is a collection of smooth *linear* vector fields satisfying Hörmander's condition in Definition 1 (see also [**29**], [**1**], [**2**] and [**13**]). Thus the degeneracies are incorporated only in the linear part of the operator, i.e. the vector fields X_j, while the nonlinearities occur only in the elliptic part of the operator, i.e. the matrix $M(x, \mathbf{p})$. In the above application to the Monge-Ampère equation, the vector fields degenerate nonlinearly, and these methods do not apply.

CHAPTER 2

Comparisons of conditions

In this section, we will compare our flag condition in Definition 13 to Hörmander's commutation condition in Definition 1, showing that they coincide for real-analytic diagonal vector fields, and compare our homogeneous space structures used in the proofs of Theorems 17 and 20 to the metric space of subunit balls. Proposition 44 provides a vital link in one of our proofs of Theorem 17. The remaining results of this section provide useful perspective, but not all of them will be used in the sequel.

1. Flags and commutators

We begin by demonstrating here the equivalence of Hörmander's commutation condition in Definition 1 and the flag condition in Definition 13 when the diagonal vector fields $X_j = a_j(x) \frac{\partial}{\partial x_j}$, $1 \le j \le n$, are real-analytic.

PROPOSITION 29. *If $a_j(x)$ is real-analytic in Ω for $1 \le j \le n$, then the vector fields $X_j = a_j(x) \frac{\partial}{\partial x_j}$, $1 \le j \le n$, satisfy Definition 1 if and only if they satisfy Definition 13 in Ω.*

PROOF. Suppose first that $\{X_j\}_{j=1}^n$ satisfies Definition 13 in Ω. Fix $x \in \Omega$ and let $\mathcal{W}$ denote the linear span of $\{X_j\}_{j=1}^n$ and their commutators of all orders at the point x. We must show that $\mathcal{W} = \mathbb{R}^n$. Since $i \in \mathcal{I}_1$ implies that $X_i(x) \ne 0$, we have that $\mathcal{V}_1 \subset \mathcal{W}$. Now fix $i \in \mathcal{I}_2 \backslash \mathcal{I}_1$. Then a_i is a real-analytic function that is not identically zero on $x + \mathcal{V}_1$, and thus we must have $\frac{\partial^{|\alpha|}}{\partial x_1^{\alpha_1} \dots \partial x_n^{\alpha_n}} a_i(x) \ne 0$ for some multi-index α of minimal length with $\alpha_j = 0$ for $j \notin \mathcal{I}_1$. But as we will show, this implies that the direction $\mathbf{e}_i$ is in the span $\mathcal{W}_i$ of $\{X_i\} \cup \{X_j\}_{j \in \mathcal{I}_1}$ and their commutators up to order $|\alpha|$ at the point x (note that we identify $\mathbf{e}_i$ with $\frac{\partial}{\partial x_i}$). We proceed by induction on the length $|\alpha|$ of α, remembering that $|\alpha|$ is minimal. If $|\alpha| = 0$, then $a_i(x) \ne 0$ and so $\frac{\partial}{\partial x_i} \in \mathcal{W}_i$. If $|\alpha| = 1$, say $\alpha = \mathbf{e}_j$ with $j \in \mathcal{I}_1$, then $a_i(x) = 0$, $a_j(x) \ne 0$ and we compute that

$$(53) \qquad [X_j, X_i] = X_j X_i - X_i X_j = a_j \frac{\partial a_i}{\partial x_j} \frac{\partial}{\partial x_i} - a_i \frac{\partial a_j}{\partial x_i} \frac{\partial}{\partial x_j}$$

equals $a_j(x) \frac{\partial a_i}{\partial x_j}(x) \frac{\partial}{\partial x_i}$ at x. Since $\frac{\partial a_i}{\partial x_j}(x) \ne 0$ by assumption, we have that $\frac{\partial}{\partial x_i} \in \mathcal{W}_i$ since $a_j(x) \ne 0$ as noted above. Now if $|\alpha| = 2$, say $\alpha = \mathbf{e}_j + \mathbf{e}_k$ with $j, k \in \mathcal{I}_1$, then $a_i(x) = \frac{\partial a_i}{\partial x_j}(x) = \frac{\partial a_i}{\partial x_k}(x) = 0$, $a_j(x) \ne 0$, $a_k(x) \ne 0$ and we compute that

$$
\begin{aligned}
[X_k, [X_j, X_i]] &= X_k [X_j, X_i] - [X_j, X_i] X_k \\
&= a_k \left\{ \frac{\partial}{\partial x_k} \left(a_j \frac{\partial a_i}{\partial x_j} \right) \right\} \frac{\partial}{\partial x_i} - a_k \left\{ \frac{\partial}{\partial x_k} \left(a_i \frac{\partial a_j}{\partial x_i} \right) \right\} \frac{\partial}{\partial x_j} \\
&\quad - \left(a_j \frac{\partial a_i}{\partial x_j} \frac{\partial a_k}{\partial x_i} - a_i \frac{\partial a_j}{\partial x_i} \frac{\partial a_k}{\partial x_j} \right) \frac{\partial}{\partial x_k} \\
&= \left(a_k \frac{\partial a_j}{\partial x_k} \frac{\partial a_i}{\partial x_j} + a_k a_j \frac{\partial^2 a_i}{\partial x_k \partial x_j} \right) \frac{\partial}{\partial x_i} \quad (\mathrm{mod}\, \mathcal{V}_1)
\end{aligned}
$$

equals $a_k(x) a_j(x) \frac{\partial^2 a_i}{\partial x_k \partial x_j}(x) \frac{\partial}{\partial x_i}$ $(\mathrm{mod}\,\mathcal{V}_1)$ at x. Since $\frac{\partial^2 a_i}{\partial x_k \partial x_j}(x) \neq 0$ by assumption, we have that $\frac{\partial}{\partial x_i} \in \mathcal{W}_i$ since $a_j(x), a_k(x) \neq 0$ as noted above. Continuing inductively in this way, we obtain that $\frac{\partial}{\partial x_i} \in \mathcal{W}_i \subset \mathcal{W}$. Thus we have shown that for $i \in \mathcal{I}_2$, the directions $\mathbf{e}_i$ are in $\mathcal{W}$, or equivalently that $\mathcal{V}_2 \subset \mathcal{W}$.

Now we proceed to show that the direction $\mathbf{e}_i$ lies in $\mathcal{W}$ for $i \in \mathcal{I}_3 \backslash \mathcal{I}_2$. Again, $i \in \mathcal{I}_3 \backslash \mathcal{I}_2$ implies that a_i is a real-analytic function that is not identically zero on $x + \mathcal{V}_2$, and thus we must have $\frac{\partial^{|\alpha|}}{\partial x_1^{\alpha_1} \ldots \partial x_n^{\alpha_n}} a_i(x) \neq 0$ for some multi-index α of minimal length with $\alpha_j = 0$ for $j \notin \mathcal{I}_2$. As above, it follows by induction on the length $|\alpha|$ of α that the direction $\mathbf{e}_i$ is in the span of $\{X_i\} \cup \{X_j\}_{j \in \mathcal{I}_2}$ and their commutators up to order $|\alpha|$ at the point x. Thus for $i \in \mathcal{I}_3$, the directions $\mathbf{e}_i$ are in $\mathcal{W}$, or $\mathcal{V}_3 \subset \mathcal{W}$. Iterating this argument we eventually obtain that $\mathbb{R}^n = \mathcal{V}_m \subset \mathcal{W}$, and so the condition in Definition 1 holds as required.

Conversely, suppose that $\{X_j\}_{j=1}^n$ satisfies Definition 1. With x fixed define

$$
\begin{aligned}
\mathcal{I}_1 &= \{i : a_i(x) \neq 0\}, \\
\mathcal{V}_1 &= span\{\mathbf{e}_i : i \in \mathcal{I}_1\}.
\end{aligned}
$$

Then $\mathcal{I}_1 \neq \phi$, since otherwise $a_j(x) = 0$ for all $1 \leq j \leq n$, and it follows from (53) that all commutators of $\{X_j\}_{j=1}^n$ of finite order vanish at x as well, contradicting Definition 1. Now define

$$
\begin{aligned}
\mathcal{I}_2 &= \{i : a_i \text{ is not identically } 0 \text{ on } (x + \mathcal{V}_1) \cap \mathcal{N} \text{ for any neighbourhood } \mathcal{N} \text{ of } x\}, \\
\mathcal{V}_2 &= span\{\mathbf{e}_i : i \in \mathcal{I}_2\}.
\end{aligned}
$$

We claim that $\mathcal{I}_2 \backslash \mathcal{I}_1 \neq \phi$ unless $\mathcal{I}_1 = \{1, 2, \ldots, n\}$. If not, then $\mathcal{I}_1 \subsetneq \{1, 2, \ldots, n\}$ and $\mathcal{I}_2 \backslash \mathcal{I}_1 = \phi$. Thus there is a neighbourhood $\mathcal{N}$ of x such that all a_i with $i \notin \mathcal{I}_1$ vanish identically on $(x + \mathcal{V}_1) \cap \mathcal{N}$. This motivates the following definitions. Let $\mathcal{X}$ denote the Lie algebra generated by the vector fields $\{X_j\}_{j=1}^n$, and define the linear subspace

$$
\mathcal{Y}_1 = \left\{ T = \sum_{j=1}^n b_j \frac{\partial}{\partial x_j} \in \mathcal{X} : b_j \equiv 0 \text{ on } (x + \mathcal{V}_1) \cap \mathcal{N} \text{ for } j \notin \mathcal{I}_1 \right\}.
$$

Note that $X_i = a_i \frac{\partial}{\partial x_i} \in \mathcal{Y}_1$ for all $1 \leq i \leq n$. Thus it is enough to show that $\mathcal{Y}_1$ is a Lie algebra, since it then follows that $\mathcal{X} = \mathcal{Y}_1$, contradicting Definition 1 (note that $\mathcal{Y}_1$ does not span $\mathbb{R}^n$ at x since $\mathcal{I}_1$ is a proper subset). So let $S = \sum_{j=1}^n c_j \frac{\partial}{\partial x_j}, T =$

$\sum_{j=1}^{n} b_j \frac{\partial}{\partial x_j} \in \mathcal{Y}_1$ and compute that

$$[S,T] = ST - TS = \sum_{\ell=1}^{n} d_\ell \frac{\partial}{\partial x_\ell}$$

where

$$d_\ell = \sum_{j=1}^{n} \left\{ c_j \frac{\partial b_\ell}{\partial x_j} - b_j \frac{\partial c_\ell}{\partial x_j} \right\}.$$

To show that $[S,T] \in \mathcal{Y}_1$, it suffices to show that when $\ell \notin \mathcal{I}_1$ each product, $c_j \frac{\partial b_\ell}{\partial x_j}$ and $b_j \frac{\partial c_\ell}{\partial x_j}$, vanishes identically on $(x + \mathcal{V}_1) \cap \mathcal{N}$. If $j \in \mathcal{I}_1$, then both $\frac{\partial b_\ell}{\partial x_j}$ and $\frac{\partial c_\ell}{\partial x_j}$ vanish identically on $(x + \mathcal{V}_1) \cap \mathcal{N}$ since $\ell \notin \mathcal{I}_1$ and $S, T \in \mathcal{Y}_1$. On the other hand, if $j \notin \mathcal{I}_1$, then both c_j and b_j vanish identically on $(x + \mathcal{V}_1) \cap \mathcal{N}$. This completes the proof that $\mathcal{I}_2 \backslash \mathcal{I}_1 \neq \phi$ unless $\mathcal{I}_1 = \mathbb{R}^n$.

We now continue by defining inductively,

$$\mathcal{I}_{j+1} = \{i : a_i \text{ is not identically } 0 \text{ on } (x + \mathcal{V}_j) \cap \mathcal{N} \text{ for any neighbourhood } \mathcal{N} \text{ of } x\},$$
$$\mathcal{V}_{j+1} = span\{\mathbf{e}_i : i \in \mathcal{I}_{j+1}\}.$$

At each stage of the induction, it follows that $\mathcal{I}_{j+1} \backslash \mathcal{I}_j \neq \phi$ unless $\mathcal{I}_j = \{1, 2, ..., n\}$ using Definition 1 and the fact that for any neighbourhood $\mathcal{N}$ of x,

$$\mathcal{Y}_j = \left\{ T = \sum_{\ell=1}^{n} b_\ell \frac{\partial}{\partial x_\ell} \in \mathcal{X} : b_\ell \equiv 0 \text{ on } (x + \mathcal{V}_j) \cap \mathcal{N} \text{ for } \ell \notin \mathcal{I}_j \right\}$$

is a Lie algebra. Eventually, we reach $\mathcal{V}_m = \mathbb{R}^n$ for some $m \in \mathbb{N}$, and this shows that the equivalent formulation of Definition 13 given in Remark 14 holds.

REMARK 30. *One can easily check that the above proof establishes the stronger assertions: If the vector fields $X_j = a_j(x) \frac{\partial}{\partial x_j}$, $1 \leq j \leq n$, are smooth and satisfy Definition 13, and if $a_j(x)$ satisfies the reverse Hölder condition in each variable x_i with $j \in \mathcal{I}_{\ell+1} \backslash \mathcal{I}_\ell$ and $i \in \mathcal{I}_\ell$, uniformly in the remaining variables, then the vector fields $\{X_j\}_{j=1}^{n}$ satisfy Definition 1. This uses the fact that a smooth function f that is reverse Hölder in each variable separately, and not identically zero, is of finite type, i.e. for every x there is a multi-index α such that $\frac{\partial^{|\alpha|}}{\partial x_1^{\alpha_1} ... \partial x_n^{\alpha_n}} f(x) \neq 0$. See e.g. Guan and the first author [**17**]. Conversely, the proof as given shows that if the $a_j(x)$ are merely smooth and the vector fields $X_j = a_j(x) \frac{\partial}{\partial x_j}$, $1 \leq j \leq n$, satisfy Definition 1, then they satisfy Definition 13. Note also that part of the construction in the second half of the proof above provides a means of creating a flag of minimal length for vector fields $\{X_j\}_{j=1}^{n}$ satisfying Definition 13, even when the coefficients a_j are not real-analytic, but merely continuous. The argument also yields the uniqueness of the minimal flag.*

2. Homogeneous and prehomogeneous spaces

We now introduce the notions from the theory of homogeneous spaces that we will need. Let d be a quasimetric on an open subset Ω of $\mathbb{R}^n$, by which we mean a finite nonnegative function d on $\Omega \times \Omega$ for which there is a positive constant κ so that

(54)
$$d(x,y) = 0 \Longleftrightarrow x = y$$
$$d(x,y) \leq \kappa(d(x,z) + d(y,z))$$

for all x, y, z in Ω. A quasimetric is not in general symmetric, but see Remark 36 below. We define the ball centered at x with radius r by

$$(55) \qquad B(x, r) = \{y \in \Omega : d(x, y) < r\}, \qquad 0 < r < \infty.$$

If $d(x, y)$ is upper semicontinuous in the second variable y, for each x, then the balls $B(x, r)$ are open for $x \in \Omega$ and $r > 0$. In general the balls satisfy the engulfing property: there is a constant $\gamma > 1$ such that

$$(56) \qquad B(x, r) \cap B(y, r) \neq \phi \Longrightarrow B(y, r) \subset B(x, \gamma r).$$

Indeed, following the development in Chapter 1 of Stein [**42**], if $w \in B(y, r)$ and $z \in B(x, r) \cap B(y, r)$, then

$$
\begin{aligned}
d(x, w) \;&\leq\; \kappa\left[d(x, y) + d(w, y)\right] \\
&\leq\; \kappa\left[\kappa\left(d(x, z) + d(y, z)\right) + \kappa\left(d(w, w) + d(y, w)\right)\right] \\
&<\; \kappa\left[\kappa(r + r) + \kappa(0 + r)\right] = 3\kappa^2 r,
\end{aligned}
$$

which yields (56) with $\gamma = 3\kappa^2$. We have as well the monotonicity and scale properties

$$(57) \qquad \cup_{0 < r < s} B(x, r) = B(x, s), \qquad \text{for } s > 0, \; x \in \Omega,$$

$$(58) \qquad \cap_{r > 0} B(x, r) = \{x\} \text{ and } \cup_{r > 0} B(x, r) = \Omega, \qquad x \in \Omega.$$

Conversely, given a collection of open subsets $\{B(x, r)\}_{x \in \Omega, 0 < r < \infty}$ of Ω satisfying (56), (57) and (58), the function

$$(59) \qquad d(x, y) = \inf\{r > 0 : y \in B(x, r)\}$$

is a quasimetric on Ω that is upper semicontinuous in the second variable, and satisfies both (54) and (55). This is essentially stated in [**42**] and [**33**], but for the sake of completeness, we give here a statement and proof, as well as an extension to balls that fail to satisfy the monotonicity condition (57).

LEMMA 31. *Suppose Ω is an open subset of $\mathbb{R}^n$, and $\{B(x, r)\}_{x \in \Omega, 0 < r < \infty}$ is a family of open subsets of Ω satisfying (56), (57) and (58). Then the function d defined in (59) is upper semicontinuous in the second variable and satisfies (54) and (55). If however, the monotonicity condition (57) is relaxed to the weak monotonicity condition,*

$$(60) \qquad B(x, r) \subset B(x, s), \qquad \text{for } 0 < r \leq cs, \; x \in \Omega,$$

for some $0 < c < 1$, then d in (59) still satisfies (54) and is upper semicontinuous in the second variable, but (55) must be replaced with the equivalence condition

$$(61) \qquad B(x, r) \subset \{y \in \Omega : d(x, y) < t\} \subset B\left(x, \frac{t}{c}\right), \qquad 0 < r < t < \infty.$$

PROOF. First, d is defined and finite on $\Omega \times \Omega$ since $\cup_{r > 0} B(x, r) = \Omega$ for all $x \in \Omega$ by (58). The upper semicontinuity of d in the second variable follows from the assumption that the balls $B(x, r)$ are open.

If $d(x, y) = 0$, then $y \in \cap_{r > 0} B(x, r) = \{x\}$ by (57) and (58), even by (60) and (58), and so $x = y$. Conversely, if $x = y$, then $d(x, y) = 0$ by (59) and the first part of (58) and this proves the first assertion in (54). Now fix x, y, z and choose $r > \max\{d(x, z), d(y, z)\}$. Then $B(x, r) \cap B(y, r) \supset \{z\} \neq \phi$ implies that

$y \in B(y, r) \subset B(x, \gamma r)$ by (56). Thus $d(x, y) \leq \gamma r$ and taking the infimum over $r > \max\{d(x, z), d(y, z)\}$ yields

$$d(x, y) \leq \gamma \max\{d(x, z), d(y, z)\} \leq \gamma(d(x, z) + d(y, z)),$$

which is the second assertion in (54) with $\kappa = \gamma$.

The forward inclusion $\subset$ in (55) holds since $y \in B(x, r)$ implies $y \in B(x, t)$ for some $t < r$ by the backward inclusion of (57), which yields $d(x, y) < r$. The backward inclusion $\supset$ in (55) follows since if $d(x, y) < r$, then $y \in B(x, r)$ by the forward inclusion in (57). With the weak monotonicity condition (60) in place of (57), we obtain only the containments in (61). This completes the proof of Lemma 31.

DEFINITION 32. *A pair (Ω, d) where Ω is an open subset of $\mathbb{R}^n$ and d is a quasimetric that is upper semicontinuous in the second variable, is a* homogeneous space *if the balls $B(x, r)$ defined in (55) satisfy a doubling condition: there is a positive constant C_{doub} such that*

(62)
$$|B(x, \gamma r)| \leq C_{doub}|B(x, r)|$$

for all $x \in \Omega$ and $r > 0$, where $\gamma > 1$ is the engulfing constant in (56).

We shall also need the notion of homogeneous space without the balls being open.

DEFINITION 33. *A pair (Ω, d) where Ω is an open subset of $\mathbb{R}^n$ and $d(x, y)$ is a quasimetric that is Lebesgue measurable in the second variable y for each x, and whose balls as defined in (55) satisfy (62), is a* general homogeneous space.

Note that the doubling condition implies that the balls $B(x, r)$ in either definition above are nonempty, and in fact have positive measure, since for any fixed $r > 0$,

$$0 < |\Omega| = \sup_{m \geq 1}|B(x, \gamma^m r)| \leq \sup_{m \geq 1} C_{doub}^m |B(x, r)|$$

by (58) and (57), even by (58) and (60), and therefore $|B(x, r)| > 0$. Thus homogeneous spaces (Ω, d) on an open set Ω are characterized by a pair $(\Omega, \mathcal{B})$ where $\mathcal{B} = \{B(x, r)\}_{x \in \Omega, 0 < r < \infty}$ is a family of nonempty open subsets of Ω satisfying (56), (57), (58) and (62).

In proving Theorems 17 and 20 in section 4, we will need the following notion of prehomogeneous space, weaker than that of homogeneous space, which requires only weak monotonicity of the open sets $B(x, r)$.

DEFINITION 34. *A pair $(\Omega, \mathcal{B})$ where Ω is an open subset of $\mathbb{R}^n$ and*

$$\mathcal{B} = \{B(x, r)\}_{x \in \Omega, 0 < r < \infty}$$

is a family of nonempty open subsets of Ω, is a prehomogeneous space *if (56), (58), (62) and (60) hold.*

We often refer to the open sets $B(x, r)$ in a prehomogeneous space as preballs.

REMARK 35. *Lemma 31 shows that a prehomogeneous space $(\Omega, \mathcal{B})$ with*

$$\mathcal{B} = \{B(x, r)\}_{x \in \Omega, 0 < r < \infty}$$

is equivalent to the homogeneous space with quasimetric d given in (59). The quasi-metric balls $\{B^*(x,r)\}_{x\in\Omega,0<r<\infty}$ where $B^*(x,r) = \{y \in \Omega : d(x,y) < r\}$ satisfy

$$B(x,r) \subset B^*(x,t) \subset B\left(x,\frac{t}{c}\right), \qquad 0 < r < t < \infty.$$

We will refer to this quasimetric d as the quasimetric associated to the prehomogeneous space.

REMARK 36. *Setting $z = x$ in (54) yields $d(x,y) \leq \kappa d(y,x)$, from which we obtain that d is equivalent to the symmetric function*

(63) $$d_{sym}(x,y) = \frac{1}{2}[d(x,y) + d(y,x)].$$

The function d_{sym} satisfies (54) with a larger constant, but in general fails to be upper semicontinuous, even measurable, in the second variable (e.g. take $d(x,y) = \theta(x)|x-y|$ where $1 \leq \theta(x) \leq 2$ and θ is not Lebesgue measurable).

We now show that a homogenous space is always equivalent to a symmetric general homogeneous space.

LEMMA 37. *Given a quasimetric $d(x,y)$ that is upper semicontinuous in the second variable, there is a symmetric quasimetric $d^*_{sym}(x,y)$ that is equivalent to $d(x,y)$ and is Borel measurable in each variable separately.*

PROOF. Define
$$d^*(x,y) = \lim\inf_{z\to x} d(z,y).$$
Then d^* is lower semicontinuous in the first variable x for each y. Moreover, $d^*(x,y) = \lim_{\delta\to 0} f_\delta^x(y)$ is a monotone increasing limit of functions $f_\delta^x(y) = \inf_{|x-z|<\delta} d(z,y)$ that are upper semicontinuous in the second variable y for each x. Thus d^* is Borel measurable in each variable separately, and so then is the symmetric function
$$d^*_{sym}(x,y) = \frac{1}{2}[d^*(x,y) + d^*(y,x)].$$
It only remains to prove the equivalence of d and d^*_{sym}, which in turn is implied by the equivalence of d and d^*. Clearly $d^*(x,y) = 0$ if $d(x,y) = 0$. Fix x and y with $d(x,y) \neq 0$. By the upper semicontinuity of d in the second variable, there is $\delta_0 > 0$ such that
$$D(x,\delta_0) \subset \left\{z : d(x,z) < \alpha \equiv \frac{d(x,y)}{2\kappa}\right\}.$$
Then for $\delta < \delta_0$ and $|x-z| < \delta$, we have $d(x,z) < \alpha$ and so
$$\begin{aligned}
d(x,y) &\leq \kappa[d(x,z) + d(y,z)] \\
&\leq \kappa[\alpha + \kappa d(z,y)] \\
&= \frac{1}{2}d(x,y) + \kappa^2 d(z,y)
\end{aligned}$$
implies $d(x,y) \leq 2\kappa^2 d(z,y)$. Conversely, for $|x-z| < \delta$,
$$\begin{aligned}
d(z,y) &\leq \kappa[d(z,x) + d(y,x)] \\
&\leq \kappa[\kappa d(x,z) + \kappa d(x,y)] \\
&\leq \frac{\kappa}{2}d(x,y) + \kappa^2 d(x,y)
\end{aligned}$$

implies $d(z, y) \leq \left(\frac{\kappa}{2} + \kappa^2\right) d(x, y)$. Altogether, we have shown that

$$\frac{1}{2\kappa^2} d(x, y) \leq d(z, y) \leq \left(\frac{\kappa}{2} + \kappa^2\right) d(x, y), \qquad \text{for } |x - z| < \delta < \delta_0,$$

which yields

$$\frac{1}{2\kappa^2} d(x, y) \leq f_\delta^x(y) \leq \left(\frac{\kappa}{2} + \kappa^2\right) d(x, y), \qquad \text{for } \delta < \delta_0.$$

Now let $\delta \to 0$ to obtain

$$\frac{1}{2\kappa^2} d(x, y) \leq d^*(x, y) \leq \left(\frac{\kappa}{2} + \kappa^2\right) d(x, y),$$

which completes the proof.

We close this section by introducing the *restriction* of a prehomogeneous space $\mathbb{B} = (\Omega, \mathcal{B})$ with $\mathcal{B} = \{B(x, r)\}_{x \in \Omega, 0 < r < \infty}$ to one of its preballs $B(x_0, r_0)$, by simply intersecting the balls $B(x, r)$ with $B(x_0, r_0)$ itself, for $x \in B(x_0, r_0)$, to obtain the family

(64) $\mathcal{B}_0 = \{B_0(x, r)\}_{x \in B(x_0, r_0), 0 < r < \infty}$, where $B_0(x, r) = B(x, r) \cap B(x_0, r_0)$.

LEMMA 38. *Let* $\mathbb{B} = (\Omega, \mathcal{B})$ *with* $\mathcal{B} = \{B(x, r)\}_{x \in \Omega, 0 < r < \infty}$ *be a prehomogeneous space on* Ω, *and let* $x_0 \in \Omega$ *and* $0 < r_0 < \infty$. *Then the pair* $\mathbb{B}_0 = (B(x_0, r_0), \mathcal{B}_0)$, *where* $\mathcal{B}_0$ *is given by* (64), *is a prehomogeneous space on* $B(x_0, r_0)$ *provided that there is* $C_1 > 0$ *such that*

(65) $|B(x, r)| \leq C_1 |B(x, r) \cap B(x_0, r_0)|, \qquad for\ 0 < r \leq \frac{1}{2} r_0, x \in B(x_0, r_0).$

We refer to $\mathbb{B}_0$ as the *restriction* of $\mathbb{B}$ to its preball $B(x_0, r_0)$.

PROOF. Properties (56), (58) and (60) are immediate for $\mathbb{B}_0 = (B(x_0, r_0), \mathcal{B}_0)$. To show the doubling property (62) for $\mathbb{B}_0$, let c be as in (60) and choose $m > 1$ so large that $\gamma^{1-m} < \frac{1}{2} c^2$. Then (56) and (60) for $\mathbb{B}$, together with (65), yield (62) for $\mathbb{B}_0$ as follows. Let $x \in B(x_0, r_0)$ and $0 < r_0 < \infty$. By (64) and $m + 1$ applications of doubling (62) for the preballs balls $B(x, t)$ of $\mathbb{B}$, we have

$$|B_0(x, \gamma r)| \leq |B(x, \gamma r)| \leq C_{doub}^{m+1} |B(x, \gamma^{-m} r)|,$$

and using $\gamma^{-m} r < c\left(\frac{1}{2}\gamma^{-1} cr\right)$ in (60) gives

$$C_{doub}^{m+1} |B(x, \gamma^{-m} r)| \leq C_{doub}^{m+1} \left| B\left(x, \frac{1}{2}\gamma^{-1} cr\right) \right|.$$

Thus in the case $0 < r \leq \gamma c^{-1} r_0$, we can apply (65) using $\frac{1}{2}\gamma^{-1} cr \leq \frac{1}{2} r_0$ to obtain

$$C_{doub}^{m+1} \left| B\left(x, \frac{1}{2}\gamma^{-1} cr\right) \right| \leq C_{doub}^{m+1} C_1 \left| B\left(x, \frac{1}{2}\gamma^{-1} cr\right) \cap B(x_0, r_0) \right|.$$

Combining estimates yields

$$
\begin{aligned}
(66) \qquad |B_0(x, \gamma r)| & \leq C_{doub}^{m+1} C_1 \left| B\left(x, \frac{1}{2}\gamma^{-1} cr\right) \cap B(x_0, r_0) \right| \\
& \leq C_{doub}^{m+1} C_1 |B(x, r) \cap B(x_0, r_0)| \\
& = C_{doub}^{m+1} C_1 |B_0(x, r)|,
\end{aligned}
$$

for $0 < r \leq \gamma c^{-1} r_0$, upon using another application of (60) with $\frac{1}{2}\gamma^{-1} cr < cr$. On the other hand, for $\gamma c^{-1} r_0 < r < \infty$ we have

$$(67) \qquad |B_0(x, \gamma r)| \leq |B(x_0, r_0)| = |B(x, r) \cap B(x_0, r_0)|,$$

since by the engulfing property (56), together with (60) and $\gamma r_0 < cr$, we have $B(x_0, r_0) \subset B(x, \gamma r_0) \subset B(x, r)$. Combining (66) and (67) yields (62) for the sets in (64), and this completes the proof of Lemma 38.

There is an analogous result for the restriction of a homogeneous space $\mathbb{B} = (\Omega, \mathcal{B})$ with $\mathcal{B} = \{B(x, r)\}_{x \in \Omega, 0 < r < \infty}$ to one of its balls $B(x_0, r_0)$, $x_0 \in \Omega$ and $0 < r_0 < \infty$.

2.1. Local and extendible spaces. In our construction of the noninterference balls $A(x, r)$ in (38), and also in the construction of the flag balls $B(x, r)$ in (88) below (actually preballs in our terminology since monotonicity may fail for flag balls), we only define the balls for $x \in \Omega$ and $0 < r < \delta\, dist(x, \partial\Omega)$ for some $\delta > 0$. Recall that $dist(x, \partial\Omega)$ denotes the *Euclidean* distance from x to $\partial\Omega$ as in the Convention in subsection 1.1 of the introduction. Nevertheless, in order to take advantage of the machinery in the subsection above, we need to define the balls for all $0 < r < \infty$, while retaining all pertinent properties in the extension. To accomplish this effectively, we first introduce the concept of a δ-local prehomogeneous space, and the definition of an extendible δ-local prehomogeneous space. Then we show that every δ-local prehomogeneous space is locally extendible to a prehomogeneous space. The proof of this result is rather technical, and is not needed until subsection 4.5 on reducing the proofs of our extensions of Hörmander's theorem.

DEFINITION 39. *A pair $(\Omega, \mathcal{B})$ where Ω is an open subset of $\mathbb{R}^n$ and*

$$\mathcal{B} = \{B(x, r)\}_{x \in \Omega, 0 < r < \delta\, dist(x, \partial\Omega)}$$

is a family of nonempty open subsets of Ω for some $\delta > 0$, is a δ-local prehomogeneous space if the following δ-local analogues of (56), (58), (60) and (62) hold:

$$(68) \qquad B(x, r) \cap B(y, r) \neq \phi \Longrightarrow B(y, r) \subset B(x, \gamma r), \quad x, y \in \Omega,$$
$$0 < r < \delta\, dist(y, \partial\Omega), 0 < \gamma r < \delta\, dist(x, \partial\Omega);$$

$$(69) \qquad \cap_{0 < r < \delta\, dist(x, \partial\Omega)} B(x, r) = \{x\}, \quad x \in \Omega;$$

$$(70) \qquad B(x, r) \subset B(x, s), \quad x \in \Omega, \text{ for } 0 < r \leq cs < s < \delta\, dist(x, \partial\Omega);$$

$$(71) \qquad |B(x, \gamma r)| \leq C_{doub} |B(x, r)|, \quad x \in \Omega, 0 < \gamma r < \delta\, dist(x, \partial\Omega).$$

It is convenient to refer to the open sets $B(x, r)$ with $x \in \Omega$ and $0 < r < \delta\, dist(x, \partial\Omega)$ as δ-*local* preballs. We also define a δ-*local* homogeneous space in the analogous way using in place of the mononicity condition (57), the δ-local mononicity condition,

$$(72) \qquad \cup_{0 < r < s} B(x, r) = B(x, s), \quad x \in \Omega, 0 < s < \delta\, dist(x, \partial\Omega).$$

All of the results in this subsection have obvious variants for a δ-local homogeneous space, but to save notation, we will not explicitly point them out. In order to help clarify matters, we will use $\mathbb{B}$ to denote a δ-local prehomogeneous space and $\mathbb{H}$ to denote a prehomogeneous space for the remainder of this subsection.

We observe that a δ-local prehomogeneous space $\mathbb{B} = (\Omega, \mathcal{B})$ with

$$\mathcal{B} = \{B(x, r)\}_{x \in \Omega, 0 < r < \delta\ dist(x, \partial\Omega)}$$

on Ω *induces* a δ_0-local prehomogeneous space $\mathbb{B}_0 = (\Omega_0, \mathcal{B}_0)$ with

$$\mathcal{B}_0 = \{B_0(x, r)\}_{x \in \Omega_0, 0 < r < \delta_0\ dist(x, \partial\Omega_0)}$$

on Ω_0 for any open subset Ω_0 of Ω and any positive δ_0 by restricting x and r, i.e.

$$B_0(x, r) = B(x, r), \qquad x \in \Omega_0, 0 < r < \delta_0\ dist(x, \partial\Omega_0),$$

provided the following two conditions hold:

$$\text{(73)} \qquad \delta_0\ dist(x, \partial\Omega_0) \ \leq \ \delta\ dist(x, \partial\Omega) \qquad \text{for } x \in \Omega_0,$$
$$B(x, r) \ \subset \ \Omega_0, \qquad \text{for } x \in \Omega_0, 0 < r < \delta_0\ dist(x, \partial\Omega_0).$$

In addition, a prehomogeneous space $\mathbb{H} = (\Omega, \mathcal{B})$ on Ω induces a δ-local prehomogeneous space on Ω for any $0 < \delta < \infty$ simply by restricting r. Unfortunately, not all δ-local prehomogeneous spaces arise in this useful fashion, i.e. not all δ-local prehomogeneous spaces $\mathbb{B} = (\Omega, \mathcal{B})$ with $\mathcal{B} = \{B(x, r)\}_{x \in \Omega, 0 < r < \delta\ dist(x, \partial\Omega)}$ can have preballs defined for $x \in \Omega$ and $r \geq \delta\ dist(x, \partial\Omega)$ as well in such a way that the resulting collection of open sets $\{B(x, r)\}_{x \in \Omega, 0 < r < \infty}$ forms a prehomogeneous space on Ω.

DEFINITION 40. *A δ-local prehomogeneous space $\mathbb{B} = (\Omega, \mathcal{B})$ with*

$$\mathcal{B} = \{B(x, r)\}_{x \in \Omega, 0 < r < \delta dist(x, \partial\Omega)}$$

on Ω is said to be extendible *if there is a prehomogeneous space $\mathbb{H}^* = (\Omega, \mathcal{B}^*)$ with*

$$\mathcal{B}^* = \{B^*(x, r)\}_{x \in \Omega, 0 < r < \infty}$$

on Ω such that

$$B^*(x, r) = B(x, r), \qquad x \in \Omega, 0 < r < \delta\ dist(x, \partial\Omega).$$

Our main result in this subsection is that a δ-local prehomogeneous space on Ω is "locally" extendible, in a sense to be made precise below, provided our standing assumption

$$\text{(74)} \qquad B(x, r) \subset D(x, C_{euc}r), \qquad 0 < r < \delta\ dist(x, \partial\Omega),$$

is in force, δ satisfies

$$\text{(75)} \qquad \delta < C_{euc}^{-1} \min\left\{\frac{1}{2}, \gamma(\gamma - 1)\right\},$$

where γ is as in (68), and the following "relative proportion" condition holds:
(76)
$$|B(x, r)| \leq C_1 |B(x, r) \cap B(y, s)|, \text{ for } y \in \Omega, x \in B(y, s), 0 < 2r \leq s < \delta\ dist(y, \partial\Omega).$$

Note that if δ satisfies (75), then the δ-local preball $B(x, r)$ in (76) is defined since $0 < r < \delta\ dist(x, \partial\Omega)$ then holds automatically. Indeed,

$$\delta\ dist(x, \partial\Omega) \geq \delta\ dist(y, \partial\Omega) - \delta|x - y| \geq s - \delta C_{euc}s > \frac{s}{2},$$

since $\delta < C_{euc}^{-1}\frac{1}{2}$, and so $r \leq \frac{s}{2} < \delta\ dist(x, \partial\Omega)$.

PROPOSITION 41. *Let* $\mathbb{B} = (\Omega, \mathcal{B})$ *with* $\mathcal{B} = \{B(x,r)\}_{x\in\Omega, 0<r<\delta\, dist(x,\partial\Omega)}$ *be a* δ-*local prehomogeneous space on* Ω *satisfying (74), (75) and (76). Then for every* $x_0 \in \Omega$, *there is an open subset* $\Omega_0 \subset \Omega$ *with* $x_0 \in \Omega_0$ *and a positive number* δ_0 *satisfying (73) such that the* δ_0-*local prehomogeneous space* $\mathbb{B}_0$ *induced on* Ω_0 *by* $\mathbb{B}$ *is extendible.*

We will use Lemma 38, as well as the following definition and lemma, in the proof of Proposition 41.

DEFINITION 42. *Let* $\mathbb{B} = (\Omega, \mathcal{B})$ *with* $\mathcal{B} = \{B(x,r)\}_{x\in\Omega, 0<r<\delta\, dist(x,\partial\Omega)}$ *be a* δ-*local prehomogeneous space on* Ω *satisfying (74) and (75). Given* $x_0 \in \Omega$ *and a number* r_0 *such that*

$$(77) \qquad\qquad 0 < \gamma^2 r_0 < \delta\, dist(x_0, \partial\Omega),$$

we define a localized collection

$$\mathcal{B}^*(x_0, r_0) = \{B^*(x,r)\}_{x\in B(x_0,r_0), 0<r<\infty}$$

of open subsets of $B(x_0, r_0)$ *by*

$$(78)\quad B^*(x,r) = \begin{cases} B(x,r) \cap B(x_0, r_0) & \text{if } x \in B(x_0, r_0), \quad 0 < r \leq \gamma r_0 \\ B(x_0, r_0) & \text{if } x \in B(x_0, r_0), \quad r \geq \gamma r_0 \end{cases}.$$

Note that $B^*(x,r)$ in (78) of Definition 42 is well-defined. To see this, we first claim that if $x \in B(x_0, r_0)$ and $0 < r \leq \gamma r_0$, then $0 < r < \delta\, dist(x, \partial\Omega)$ so that $B(x,r)$ is defined. Indeed, $B(x_0, r_0) \subset D(x_0, C_{euc}r_0)$ by (74), and so if $x \in B(x_0, r_0)$, then $|x - x_0| < C_{euc}r_0$. Thus by (75) and (77),

$$(79) \qquad \delta\, dist(x, \partial\Omega) \geq \delta\, dist(x_0, \partial\Omega) - \delta|x - x_0| > \gamma^2 r_0 - \delta C_{euc}r_0 > \gamma r_0,$$

since $\delta < C_{euc}^{-1}\gamma(\gamma - 1)$, which proves our claim. Of equal importance is the fact that $B(x_0, r_0) \subset B(x, \gamma r_0)$ if $x \in B(x_0, r_0)$, which follows from (68) since $x \in B(x_0, r_0) \cap B(x, r_0)$, and where the conditions required by (68) follow from $0 < \gamma^2 r_0 < \delta$ and (79). Thus for $r = \gamma r_0$, the two definitions in (78) coincide, and this completes the proof that $B^*(x,r)$ in (78) is well-defined.

LEMMA 43. *Let* Ω *be an open subset of* $\mathbb{R}^n$, $\delta > 0$ *satisfy (75) and let* $\mathbb{B} = (\Omega, \mathcal{B})$ *with* $\mathcal{B} = \{B(x,r)\}_{x\in\Omega, 0<r<\delta\, dist(x,\partial\Omega)}$ *be a* δ-*local prehomogeneous space on* Ω *satisfying (74) and (76). Fix* $x_0 \in \Omega$ *and* r_0 *satisfying (77). Let* $\Omega^* = B(x_0, r_0)$ *and* $\Omega_0 = B(x_0, c\gamma^{-1}r_0)$ *where* c *is as in (70). Then* $\mathbb{H}^* = (\Omega^*, \mathcal{B}^*(x_0, r_0))$, *with* $\mathcal{B}^*$ *as in Definition 42, is a prehomogeneous space on* Ω^*,

$$\Omega_0 = B^*(x_0, c\gamma^{-1}r_0),$$

and we have

$$(80) \qquad B^*(x,r) = B(x,r), \quad \text{for } x \in \Omega_0, 0 < r < c\gamma^{-1}r_0,$$

and also the analogue of (65) for $\mathbb{H}^*$ *and its preball* $\Omega_0 = B^*(x_0, c\gamma^{-1}r_0)$:

$$(81) \qquad |B^*(x,r)| \leq C|B^*(x,r) \cap B^*(x_0, c\gamma^{-1}r_0)|,$$

$$\text{for } x \in B^*(x_0, c\gamma^{-1}r_0), 0 < r \leq \frac{1}{2}c\gamma^{-1}r_0.$$

PROOF. We first show that $\mathbb{H}^*$ is a prehomogeneous space on Ω^*. It is not hard to verify the properties (56), (58), and (60) for the family $\mathcal{B}^* (x_0, r_0) = \{B^* (x, r)\}_{x \in \Omega^*, 0 < r < \infty}$ on $\Omega^* = B (x_0, r_0)$. For example, to verify the engulfing property (56), suppose $x, y \in \Omega^* = B (x_0, r_0), 0 < r < \infty$ and $B^* (x, r) \cap B^* (y, r) \neq \varnothing$. In the case $0 < r \leq r_0$, we have

$$\gamma r \leq \gamma r_0 < \delta \, dist \, (x, \partial \Omega)$$

by (79), and so from (68), we obtain $B (y, r) \subset B (x, \gamma r)$, and hence also $B^* (y, r) \subset B^* (x, \gamma r)$. On the other hand, if $r > r_0$, then $B^* (x, \gamma r) = B (x_0, r_0)$ by (78) and $B^* (yx, r) \subset B^* (x, \gamma r)$ is trivial. Property (58) is immediate and (60) follows from (70) if $s \leq \gamma r_0$, and is trivial if $s > \gamma r_0$, since then $B^* (x, s) = B (x_0, r_0)$ by (78).

We now prove the doubling property (62) for the family $\mathcal{B}^* (x_0, r_0)$ using the arguments in the proof of Lemma 38 of the previous subsection. First choose $m > 1$ so large that $\gamma^{1-m} < \min \{\frac{1}{2}, c\}$. Then for $0 < r \leq r_0$, we have from m applications of (71) using (79),

$$
\begin{aligned}
(82) \qquad |B^* (x, \gamma r)| &\leq |B (x, \gamma r)| \leq C_{doub}^m \left| B \left(x, \gamma^{1-m} r \right) \right| \\
&\leq C_{doub}^m C_1 \left| B \left(x, \gamma^{1-m} r \right) \cap B (x_0, r_0) \right|,
\end{aligned}
$$

and followed by an application of (70) using $\gamma^{1-m} r < cr$, the above is at most

$$C_{doub}^m C_1 |B (x, r) \cap B (x_0, r_0)| = C_{doub}^m C_1 |B^* (x, r)|.$$

On the other hand, if $r > r_0$, we note that $B^* (x, \gamma r) = B (x_0, r_0) = B^* (x, \gamma r_0)$ by (78). Thus from the case $r = r_0$ of (82), we obtain

$$|B^* (x, \gamma r)| = |B^* (x, \gamma r_0)| \leq C_{doub}^m C_1 \left| B \left(x, \gamma^{1-m} r_0 \right) \cap B (x_0, r_0) \right|.$$

We now consider the cases $r_0 < r \leq \gamma r_0$ and $r > \gamma r_0$ separately. In the former case, we use (70) with $\gamma^{1-m} r_0 < cr$ to obtain

$$\left| B \left(x, \gamma^{1-m} r_0 \right) \cap B (x_0, r_0) \right| \leq |B (x, r) \cap B (x_0, r_0)| = |B^* (x, r)|,$$

while in the latter case, (78) yields

$$\left| B \left(x, \gamma^{1-m} r_0 \right) \cap B (x_0, r_0) \right| \leq |B (x_0, r_0)| = |B^* (x, r)|.$$

This completes the proof of (62) for the family $\mathcal{B}^* (x_0, r_0)$, and thus establishes that $\mathbb{H}^*$ is a prehomogeneous space on Ω^*.

Since $\gamma > 1$, (70) yields $\Omega_0 = B \left(x_0, c\gamma^{-1} r_0 \right) \subset B (x_0, r_0)$ and so $\Omega_0 = B^* \left(x_0, c\gamma^{-1} r_0 \right)$.

Now we claim that if $x \in \Omega_0 = B \left(x_0, c\gamma^{-1} r_0 \right)$ and $0 < r < c\gamma^{-1} r_0$, then

$$B^* (x, r) = B (x, r) \cap B (x_0, r_0) = B (x, r).$$

Indeed,

$$(83) \qquad x \in B \left(x_0, c\gamma^{-1} r_0 \right) \subset B \left(x_0, \gamma^{-1} r_0 \right)$$

by (70), and so $B \left(x, \gamma^{-1} r_0 \right) \cap B \left(x_0, \gamma^{-1} r_0 \right)$ is not empty since it contains x. Thus (68) implies that

$$(84) \qquad B \left(x, \gamma^{-1} r_0 \right) \subset B (x_0, r_0),$$

provided the two conditions required by (68) hold. The two conditions required by (68) for this application are

$$0 < \gamma^{-1} r_0 < \delta \, dist \, (x, \partial \Omega) \text{ and } 0 < r_0 < \delta \, dist \, (x_0, \partial \Omega).$$

Now the second of these follows immediately from the hypothesis $0 < \gamma^2 r_0 < \delta\, dist\, (x_0, \partial\Omega)$, and the first follows from (79). This proves (84), and hence by (70) with $r < c\gamma^{-1} r_0$, we obtain

$$B\,(x, r) \subset B\,\left(x, \gamma^{-1} r_0\right) \subset B\,(x_0, r_0)$$

as required.

To see that $\mathcal{B}^*\,(x_0, r_0)$ and Ω_0 satisfy (81), take $y = x_0$ and $s = c\gamma^{-1} r_0$ in (76) and use (80) together with the fact that

$$s = c\gamma^{-1} r_0 < \gamma^2 r_0 < \delta\, dist\, (x_0, \partial\Omega)\,.$$

This completes the proof of Lemma 43.

PROOF (OF PROPOSITION 41). Given $x_0 \in \Omega$ and a number r_0 satisfying (77), let Ω_0 and Ω^* be as in Lemma 43, so that $\mathbb{H}^* = (\Omega^*, \mathcal{B}^*\,(x_0, r_0))$ is a prehomogeneous space on Ω^*. By Lemma 38, the restriction of $\mathbb{H}^*$ to the preball $\Omega_0 = B\,\left(x_0, c\gamma^{-1} r_0\right) = B^*\,\left(x_0, c\gamma^{-1} r_0\right)$ is a prehomogeneous space $\mathbb{H}_0^* = (\Omega_0, \mathcal{B}_0^*)$ on Ω_0 where $\mathcal{B}_0^* = \{B_0^*\,(x, r)\}_{x \in \Omega_0, 0 < r < \infty}$ and

$$B_0^*\,(x, r) = B^*\,(x, r) \cap B\,\left(x_0, c\gamma^{-1} r_0\right)\,, \qquad x \in \Omega_0, 0 < r < \infty.$$

Note that hypothesis (65) of Lemma 38 holds for $\mathbb{H}^*$ and its preball Ω_0 by (81) of Lemma 43. We now claim that if

$$(85) \qquad\qquad \delta_0 = \min\left\{C_{euc}^{-1}, \frac{c\gamma^{-1} r_0}{diam\,(\Omega_0)}\right\},$$

then (73) holds, and $\mathbb{H}_0^*$ is an extension of $\mathbb{B}_0$, the δ_0-local prehomogeneous space induced on Ω_0 by $\mathbb{B}$. Indeed, $\mathbb{B}_0 = (\Omega_0, \mathcal{B}_0)$ where $\mathcal{B}_0 = \{B_0\,(x, r)\}_{x \in \Omega_0, 0 < r < \delta_0\, dist(x, \partial\Omega_0)}$ and

$$B_0\,(x, r) = B\,(x, r)\,, \qquad x \in \Omega_0 = B\,\left(x_0, c\gamma^{-1} r_0\right), 0 < r < \delta_0\, dist\,(x, \partial\Omega_0).$$

We claim that (73) holds. Indeed, if $x \in \Omega_0$ and r satisfies $0 < r < \delta_0\, dist\,(x, \partial\Omega_0)$, we have

$$0 < r < \delta_0\, dist\,(x, \partial\Omega_0) \leq \frac{dist\,(x, \partial\Omega_0)}{diam\,(\Omega_0)} c\gamma^{-1} r_0 \leq c\gamma^{-1} r_0,$$

which when combined with (79) yields

$$0 < r < c\gamma^{-1} r_0 < \gamma r_0 < \delta\, dist\,(x, \partial\Omega)\,.$$

This establishes the first assertion in (73). Moreover, we also have $B^*\,(x, r) = B\,(x, r)$ by (80), and then from (74) that

$$B^*\,(x, r) = B\,(x, r) \subset D\,(x, C_{euc} r) \subset \Omega_0,$$

since $C_{euc} r < C_{euc} \delta_0\, dist\,(x, \partial\Omega_0) \leq dist\,(x, \partial\Omega_0)$, which is the second assertion in (73). Thus the induced space $\mathbb{B}_0$ is defined and

$$B_0^*\,(x, r) = B^*\,(x, r) \cap \Omega_0 = B\,(x, r)$$

for $x \in \Omega_0$ and $0 < r < \delta_0\, dist\,(x, \partial\Omega_0)$. This shows that $\mathbb{H}_0^*$ is an extension of $\mathbb{B}_0$, and completes the proof of Proposition 41.

2.2. Construction of the flag balls. We have already defined the family $\mathcal{A}$ of "noninterference balls" $A(x,r)$ used in the statement and proof of Theorem 20, and given in (38) as

$$A(x,r) = \prod_{j=1}^{n} (x_j - A_j(x,r), x_j + A_j(x,r))$$

where $A_j(x,r) = \int_0^r a_j(x_1 + t, x_2, ..., x_n)\, dt$ is defined when the segment joining x and $x + (r, 0, ..., 0)$ lies in Ω. We will now construct the "flag" balls $B(x,r)$ used in the proof of Theorem 17.

We assume the vector fields $\{X_j\}_{j=1}^{n} = \left\{ \frac{\partial}{\partial x_1}, a_2(x)\frac{\partial}{\partial x_2}, ..., a_n(x)\frac{\partial}{\partial x_n} \right\}$ are continuous in Ω, and that $a_j(x)$ is Lipschitz continuous in $x_2, ..., x_n$ uniformly in x_1, and reverse Hölder of infinite order in each variable x_i with $i \neq j$, uniformly in the remaining variables. We begin with some heuristics. Our goal is to use the flag condition to construct a family of open rectangles

$$B(x,r) = \prod_{j=1}^{n} (x_j - B_j(x,r), x_j + B_j(x,r))$$

for $x \in \Omega$, $0 < r < \delta\, dist(x, \partial\Omega)$, that are related to the vector fields $\{X_j\}_{j=1}^{n}$ in the sense that there are positive constants c, C such that

$$(86) \quad cB_j(x,r) \leq \sup_{z \in B(x,r)} ra_j(z) \leq CB_j(x,r), \qquad x \in \Omega, \ 0 < r < \delta\, dist(x, \partial\Omega),$$

for $1 \leq j \leq n$. Note that (86) says that the jth side length $B_j(x,r)$ of the rectangle $B(x,r)$ is comparable to r times the supremum of a_j over the rectangle $B(x,r)$. If the a_j were essentially constant, this would be the maximum distance a subunit curve could travel in the jth direction for time r, and the rectangle $B(x,r)$ would be equivalent to the Fefferman-Phong ball $K(x,r)$. For us, the importance of (86) is that it provides a key link in establishing that the rectangles $B(x,r)$ lead to a prehomogeneous space, as in Definition 34 above. The greedy algorithm we employ below actually achieves the following stronger form of (86): there are positive constants c, C such that for every $x \in \Omega$, $0 < r < \delta\, dist(x, \partial\Omega)$, there is a permutation $\{j_1, j_2, ..., j_n\}$ of $\{1, 2, ..., n\}$ with $j_1 = 1$ satisfying

$$(87) \qquad cB_{j_i}(x,r) \ \leq \ \sup_{z_{j_\ell} = x_{j_\ell}, \ell \geq i \text{ and } |z_{j_\ell} - x_{j_\ell}| \leq B_{j_\ell}(x,r), \ell < i} ra_{j_i}(z)$$

$$\leq \ \sup_{z \in B(x,r)} ra_{j_i}(z) \leq CB_{j_i}(x,r),$$

$x \in \Omega$, $0 < r < \delta\, dist(x, \partial\Omega)$ and $1 \leq i \leq n$. See Remark 91 in subsection 5.2 for a proof of (87). Since we are assuming that the a_j are reverse Hölder in x_1 uniformly in $x_2, ..., x_n$, it follows that

$$\sup_{z \in B(x,r)} ra_i(z) \text{ is essentially } \sup_{z \in B(x,r)} A_i(z,r),$$

and this motivates the use of A_i to implement our greedy algorithm, which we now describe.

We claim there is a sufficiently small $\delta > 0$, depending only on the dimension n and $C_{\max}$ in (37), in order that the following definitions make sense. Fix $x \in \Omega$

and $0 < r < \delta \, dist \, (x, \partial\Omega)$. By (39), we have

$$A_j \, (x,r) = \int_0^r a_j \, (x_1 + t, x_2, ..., x_n) \, dt,$$

$1 \leq j \leq n$, so that $A_1 \, (x,r) = r$. Now we inductively define a rearrangement $\{j_2, ..., j_n\}$ of $\{2, ..., n\}$ and nonnegative numbers $B_{j_2} \, (x,r) \, , ..., B_{j_n} \, (x,r)$ as follows: First define

$$\begin{aligned} A_{j_2} \, (x,r) &= \max_{2 \leq j \leq n} A_j \, (x,r), \\ B_{j_2} \, (x,r) &= A_{j_2} \, (x,r). \end{aligned}$$

Then for $j \neq j_2$ set

$$\Phi_j^2 \, (x,r) = \max \left\{ A_j \, (z,r) : |z_i - x_i| \leq \chi_{\{j_2\}} \, (i) \, B_i \, (x,r) \, , 1 \leq i \leq n \right\},$$

and define

$$\begin{aligned} \Phi_{j_3}^2 \, (x,r) &= \max_{j \neq j_2} \Phi_j^2 \, (x,r), \\ B_{j_3} \, (x,r) &= \Phi_{j_3}^2 \, (x,r). \end{aligned}$$

Assuming $B_{j_2} \, (x,r) \, , ..., B_{j_m} \, (x,r)$ have already been defined, then for $j \notin \{j_2, ..., j_m\}$, set

$$\Phi_j^m \, (x,r) = \max \left\{ A_j \, (z,r) : |z_i - x_i| \leq \chi_{\{j_2,...,j_m\}} \, (i) \, B_i \, (x,r) \, , 1 \leq i \leq n \right\},$$

and define

$$\begin{aligned} \Phi_{j_{m+1}}^m \, (x,r) &= \max_{j \notin \{j_2,...,j_m\}} \Phi_j^m \, (x,r), \\ B_{j_{m+1}} \, (x,r) &= \Phi_{j_{m+1}}^m \, (x,r). \end{aligned}$$

This inductively defines $B_{j_2} \, (x,r) \, , ..., B_{j_n} \, (x,r)$.

   Note: If we assume that the vector fields $\{X_j\}_{j=1}^n$ satisfy the flag condition in Definition 13, then we have the important property that $B_{j_m} \, (x,r) > 0$ for $2 \leq m \leq n$ and $r > 0$.

   We now define open rectangles

$$(88) \qquad B \, (x,r) = (x_1 - r, x_1 + r) \times \prod_{j=2}^n (x_j - B_j \, (x,r) , x_j + B_j \, (x,r)),$$

for $x \in \Omega$, $0 < r < \delta \, dist \, (x, \partial\Omega)$, which we refer to as "flag balls". Note again that if δ is sufficiently small depending on $C_{\max}$ in (37), then the rectangles $B \, (x,r)$ are well-defined and contained in Ω for $x \in \Omega$, $0 < r < \delta \, dist \, (x, \partial\Omega)$. Finally, we emphasize that the permutation $\{j_2, ..., j_n\}$ of $\{2, ..., n\}$ used to define the flag ball $B \, (x,r)$ depends on both x and r, and is analogous in spirit to the choice of N-tuple used to compute a corresponding quasimetric in Chapter 1, section 4 of [**33**].

3. Comparability with the subunit balls

   Let $X_1 = \frac{\partial}{\partial x_1}$, $X_j = a_j \, (x) \frac{\partial}{\partial x_j}$, $2 \leq j \leq n$, be a diagonal collection of continuous vector fields. We consider in this subsection the relationship of vector fields to families of sets,

$$\mathcal{B} = \{ B \, (x,r) : x \in \Omega, 0 < r < \infty \},$$

that are not necessarily balls arising from a quasimetric, and such that the sets $B \, (x,r)$ are not necessarily contained in Ω. Such families include the ones with balls $B(x,r)$ and $A(x,r)$ used in the proofs of Theorems 17 and 20, which turn out to be δ-local prehomogeneous and homogeneous spaces, as well as the subunit balls

$$\mathcal{K} = \{K(x,r) : x \in \Omega, 0 < r < \infty\}$$

with corresponding metric $\delta(x,y)$, constructed from the vector fields $\{X_j\}_{j=1}^n$ by setting $\mathcal{Q}(x,\xi) = \sum_{j=1}^n (X_j(x) \cdot \xi)^2$ in Definition 4. The family of balls $\mathcal{K}$ is generally *not* a homogeneous space on Ω.

Given a family of sets $\mathcal{B} = \{B(x,r) : x \in \Omega, 0 < r < \infty, \}$, we define a family $\widetilde{\mathcal{B}}$ of larger closed rectangles by

$$(89) \qquad B_j(x,r) = \sup_{z \in B(x,r)} |z_j - x_j|, \qquad 1 \leq j \leq n,$$

$$\widetilde{B}(x,r) = \prod_{j=1}^n [x_j - B_j(x,r), x_j + B_j(x,r)],$$

$$\widetilde{\mathcal{B}} = \left\{\widetilde{B}(x,r) : x \in \Omega, 0 < r < \infty\right\},$$

so that $B(x,r) \subset \widetilde{B}(x,r)$. The number $B_j(x,r)$ is half the jth "side length" of the smallest closed rectangular box $\widetilde{B}(x,r)$ centered at x and containing $B(x,r)$.

Given two families of sets $\mathcal{B}$ and $\mathcal{F}$ defined for $x \in \Omega$, we say that $\mathcal{B}$ is Ω-*locally contained* in $\mathcal{F}$, written $\mathcal{B} \subset \mathcal{F}$ (when Ω is understood from the context), if there are positive constants C, δ such that

$$B(x,r) \subset F(x,Cr), \qquad x \in \Omega, \, 0 < Cr < \delta \, dist(x, \partial\Omega),$$

where $B(x,r) \in \mathcal{B}$ and $F(x,Cr) \in \mathcal{F}$. We say that $\mathcal{B}$ and $\mathcal{F}$ are Ω-*locally equivalent*, and write $\mathcal{B} \cong \mathcal{F}$, if both $\mathcal{B} \subset \mathcal{F}$ and $\mathcal{F} \subset \mathcal{B}$.

It turns out that the inclusion $\mathcal{B} \subset \mathcal{K}$ is implied by the existence of the following weak subrepresentation inequality relative to the vector fields $\{X_j\}_{j=1}^n$, for Lipschitz functions f and a general homogeneous space $\mathcal{B}$ on Ω: for each $y \in \Omega$, $0 < r < \delta \, dist(y, \partial\Omega)$, there is a constant $C_{y,r}$ such that

$$(90) \quad |f(x) - C_{y,r}| \leq C \int_{B(y,C_0 r)} |\nabla_a f(z)| \frac{d(x,z)}{|B(x,d(x,z))|} dz, \qquad x \in B(y,r),$$

where $\nabla_a = \left(\frac{\partial}{\partial x_1}, a_2 \frac{\partial}{\partial x_2}, ..., a_n \frac{\partial}{\partial x_n}\right)$, and d is the quasimetric for the family $\mathcal{B}$ of d-balls $B(x,s)$ with center x and radius s. Note that the integration on the right-hand side is taken over an enlarged ball $B(y,C_0 r)$, $C_0 \geq 1$. For the reverse inclusion $\mathcal{K} \subset \mathcal{B}$, we assume the a_j are Lipschitz continuous in $x_2, ..., x_n$ and reverse Hölder of infinite order in x_1, uniformly in the other variables, and we also assume the following condition limiting the size of the vector fields $X_j = a_j \frac{\partial}{\partial x_j}$ on the sets in the family $\mathcal{B}$ (which has no special structure now),

$$(91) \quad \sup_{z \in \widetilde{B}(x,r)} A_j(z,r) \leq C' B_j(x,r), \qquad B(x,r) \in \mathcal{B}, x \in \Omega, 0 < r < \delta dist(x, \partial\Omega),$$

for $1 \leq j \leq n$, where $B_j(x,r)$ is half the jth side length of the smallest rectangular box $\widetilde{B}(x,r)$ containing $B(x,r)$, and $A_j(x,r) = \int_0^r a_j(x_1 + t, x_2, ..., x_n) dt$. Here the inclusion obtained is actually $\widetilde{\mathcal{K}} \subset \widetilde{\mathcal{B}}$. Note that the noninterference condition in Definition 18 implies that the noninterference balls $A(x,r)$ satisfy the size limiting condition (91). We will see later in Lemma 89 that the flag balls $B(x, r)$ also satisfy the size limiting condition (91) under some natural hypotheses.

PROPOSITION 44. *Suppose* $X_1 = \frac{\partial}{\partial x_1}$, $X_j = a_j(x) \frac{\partial}{\partial x_j}$, $2 \leq j \leq n$, *where the* a_j *are continuous in* Ω, *and let* $\mathcal{K}$ *be the family of subunit balls associated to the quadratic form* $Q(x, \xi) = \sum_{j=1}^{n} (X_j(x) \cdot \xi)^2$ *as in Definition 4. If the functions* a_j *are Lipschitz continuous in* $x_2, ..., x_n$ *uniformly in* x_1, *and reverse Hölder of infinite order in* x_1 *uniformly in* $x_2, ..., x_n$, *and if* $\mathcal{B}$ *is a family of sets satisfying the size limiting condition (91), then* $\widetilde{\mathcal{K}} \subset \widetilde{\mathcal{B}}$. *Conversely, if* $\mathcal{B}$ *is the family of balls in a general homogeneous space on* Ω, *and satisfies the weak subrepresentation inequality (90), then* $\mathcal{B} \subset \mathcal{K}$.

REMARK 45. *In Proposition 44, one can weaken the subrepresentation inequality (90) even further by adding to the right-hand side the average*

$$(92) \qquad Cr \frac{1}{|B(y, C_0 r)|} \int_{B(y, C_0 r)} |\nabla_a f(z)| \, dz.$$

While this generally results in a strictly weaker condition, the average (92) is in fact already dominated by the right-hand side of (90) if in addition to doubling, the balls are reverse doubling of order one:

$$(93) \qquad |B(x, r)| \leq C \frac{r}{t} |B(x, t)|, \qquad x \in \Omega, \, 0 < r < \delta \, dist(x, \partial\Omega).$$

The flag balls, noninterference balls, and subunit balls all satisfy (93) since they are weakly monotone (60) and have diameters comparable to their radii.

PROOF (OF PROPOSITION 44). We first demonstrate that $\widetilde{\mathcal{K}} \subset \widetilde{\mathcal{B}}$ if (91) holds along with the other hypotheses on a_j. Fix $x \in \Omega$ and $0 < r < \delta \, dist(x, \partial\Omega)$ with δ sufficiently small. We will show that $\widetilde{K}(x, r) \subset \widetilde{B}(x, C_0 r)$ for C_0 a sufficiently large constant independent of x and $r > 0$. Reorder variables so that $B_{j+1}(x, C_0 r) \leq B_j(x, C_0 r)$ for $2 \leq j < n$. Now a Lipschitz curve $\gamma(t) = (\gamma_j(t))_{j=1}^{n}$ is subunit

with respect to the matrix $\begin{bmatrix} 1 & 0 & \cdots & 0 \\ 0 & a_2(z)^2 & \cdots & 0 \\ \vdots & \vdots & \ddots & \vdots \\ 0 & 0 & \cdots & a_n(z)^2 \end{bmatrix}$ if $|\gamma'_j(t)| \leq \frac{1}{\sqrt{n}} a_j(\gamma(t))$

for *a.e.* t, and only if $|\gamma'_j(t)| \leq a_j(\gamma(t))$ for *a.e.* t. Thus we have

$$(94) \quad \begin{aligned} K_j(x, r) &= \sup\{|\gamma_j(t) - x_j| : 0 \leq t \leq r, \gamma(t) \text{ is subunit}, \, \gamma(0) = x\} \\ &\leq \sup\left\{ \int_0^r |\gamma'_j(s)| \, ds : \gamma(s) \text{ is subunit}, \, \gamma(0) = x \right\} \\ &\leq \sup\left\{ \int_0^r a_j(\gamma(s)) \, ds : \gamma(s) \text{ is subunit}, \, \gamma(0) = x \right\}. \end{aligned}$$

Since a_j is Lipschitz in $x_2, ..., x_n$ and reverse Hölder of infinite order in x_1, γ is subunit with $\gamma(0) = x$, and $X_1 = \frac{\partial}{\partial x_1}$ so that $K_1(x, r) = r$, we have for $2 \leq j \leq n$,

$$
\begin{aligned}
\int_0^r a_j(\gamma(s)) \, ds \quad &\leq \quad \int_0^r |a_j(\gamma_1(s), \gamma_2(s), ..., \gamma_n(s)) - a_j(\gamma_1(s), x_2, ..., x_n)| \, ds \\
&\quad + \int_0^r a_j(\gamma_1(s), x_2, ..., x_n) \, ds \\
&\leq \quad \sum_{i=2}^n r \sup_{z \in \widetilde{K}(x,r)} \left| \frac{\partial a_j}{\partial x_i}(z) \right| K_i(x, r) + r \max_{|z_1 - x_1| \leq r} a_j(z_1, x_2, ..., x_n) \\
&\leq \quad Cr \sum_{i=2}^n K_i(x, r) + C \int_0^r a_j(x_1 + t, x_2, ..., x_n) \, dt.
\end{aligned}
$$

See Remark 19 for the interpretation of $\sup_{z \in \widetilde{K}(x,r)} \left| \frac{\partial a_j}{\partial x_i} \right|$. In the final inequality above, we have used that the reverse Hölder condition implies the doubling condition. It thus follows upon taking the supremum over γ and summing in j that

$$
\begin{aligned}
\sum_{j=2}^n K_j(x, r) \quad &\leq \quad Cr \sum_{j=2}^n \sum_{i=2}^n K_i(x, r) + C \sum_{j=2}^n A_j(x, r) \\
&= \quad Cr(n-1) \sum_{i=2}^n K_i(x, r) + C \sum_{j=2}^n A_j(x, r).
\end{aligned}
$$

Now a_j is reverse Hölder of infinite order in x_1 and so $A_j(x, r)$ is reverse doubling in r: $A_j(x, r) \leq C \left(\frac{r}{t} \right)^{\widetilde{D}} A_j(x, t)$, $0 < r \leq t < \infty$. Thus for r small enough and C_0 large enough, by absorbing the first term on the right above, and then applying reverse doubling, we have

$$
\sum_{j=2}^n K_j(x, r) \leq C \sum_{j=2}^n A_j(x, r) \leq \frac{1}{(n-1)C'} \sum_{j=2}^n A_j(x, C_0 r)
$$

where C' is the constant appearing in (91). Thus using (91) with $C_0 r$ in place of r, we obtain

$$
\begin{aligned}
K_2(x, r) \leq \sum_{j=2}^n K_j(x, r) &\leq \frac{1}{(n-1)C'} \sum_{j=2}^n A_j(x, C_0 r) \\
&\leq \frac{1}{n-1} \sum_{j=2}^n B_j(x, C_0 r) \leq B_2(x, C_0 r).
\end{aligned}
$$

Recall that we reordered variables so that $B_{j+1}(x, C_0 r) \leq B_j(x, C_0 r)$ for $2 \leq j < n$.

We now proceed by induction. So assume that $K_i(x, r) \leq B_i(x, C_0 r)$ for $2 \leq i \leq \ell - 1$. Then since a_j is Lipschitz in $x_2, ..., x_n$ and reverse Hölder of infinite order in x_1, we have for $j \geq \ell$ and γ as above,

$$\int_0^r a_j\left(\gamma\left(s\right)\right) ds$$

$$\leq \int_0^r \left|a_j\left(\gamma_1\left(s\right), \gamma_2\left(s\right), ..., \gamma_n\left(s\right)\right) - a_j\left(\gamma_1\left(s\right), ..., \gamma_{\ell-1}\left(s\right), x_\ell, ..., x_n\right)\right| ds$$

$$+ \int_0^r a_j\left(\gamma_1\left(s\right), ..., \gamma_{\ell-1}\left(s\right), x_\ell, ..., x_n\right) ds$$

$$\leq \sum_{i=\ell}^n r \sup_{z \in \widetilde{K}(x,r)} \left|\frac{\partial a_j}{\partial x_i}\left(z\right)\right| K_i\left(x, r\right) + r a_j\left(\gamma_1\left(s_0\right), ..., \gamma_{\ell-1}\left(s_0\right), x_\ell, ..., x_n\right)$$

$$\leq Cr \sum_{i=\ell}^n K_i\left(x, r\right) + C \int_0^r a_j\left(x_1 + t, \gamma_2\left(s_0\right), ..., \gamma_{\ell-1}\left(s_0\right), x_\ell, ..., x_n\right) dt,$$

for some $s_0 \in [0, r]$, where we have used that $\gamma_1\left(s_0\right) \in [x_1 - r, x_1 + r]$. However, since $\left|\gamma_i\left(s_0\right) - x_i\right| \leq K_i\left(x, r\right) \leq B_i\left(x, C_0 r\right)$ for $2 \leq i \leq \ell - 1$, the reverse doubling property of $A_j\left(\left(x_1, \gamma_2\left(s_0\right), ..., \gamma_{\ell-1}\left(s_0\right), x_\ell, ..., x_n\right), r\right)$ and (91) show that for $\ell \leq j \leq n$,

$$C \int_0^r a_j\left(x_1 + t, \gamma_2\left(s_0\right), ..., \gamma_{\ell-1}\left(s_0\right), x_\ell, ..., x_n\right) dt$$

$$= C A_j\left(\left(x_1, \gamma_2\left(s_0\right), ..., \gamma_{\ell-1}\left(s_0\right), x_\ell, ..., x_n\right), r\right)$$

$$\leq \frac{1}{2nC'} A_j\left(\left(x_1, \gamma_2\left(s_0\right), ..., \gamma_{\ell-1}\left(s_0\right), x_\ell, ..., x_n\right), C_0 r\right)$$

$$\leq \frac{1}{2n} B_j\left(x, C_0 r\right),$$

for C_0 sufficiently large. Thus

$$\int_0^r a_j\left(\gamma\left(s\right)\right) ds \leq Cr \sum_{i=\ell}^n K_i\left(x, r\right) + \frac{1}{2n} B_j\left(x, C_0 r\right),$$

and it follows upon taking the supremum over γ and summing in j, that

$$\sum_{j=\ell}^n K_j\left(x, r\right) \leq Cr \sum_{j=\ell}^n \sum_{i=\ell}^n K_i\left(x, r\right) + \frac{1}{2n} \sum_{j=\ell}^n B_j\left(x, C_0 r\right).$$

For r sufficiently small, we thus obtain that

$$K_\ell\left(x, r\right) \leq \sum_{j=\ell}^n K_j\left(x, r\right) \leq \frac{1}{n} \sum_{j=\ell}^n B_j\left(x, C_0 r\right) \leq B_\ell\left(x, C_0 r\right),$$

and this completes the proof by induction.

Combining the above inequalities yields

$$\widetilde{K}\left(x, r\right) = [x_1 - r, x_1 + r] \times \prod_{j=2}^n [x_j - K_j\left(x, r\right), x_j + K_j\left(x, r\right)]$$

$$\subset [x_1 - r, x_1 + r] \times \prod_{j=2}^n [x_j - B_j\left(x, C_0 r\right), x_j + B_j\left(x, C_0 r\right)]$$

$$\subset \widetilde{B}\left(x, C_0 r\right),$$

provided $B_1\left(x, C_0 r\right) \geq r$, which follows from (91) with C_0 sufficiently large.

Conversely, we show that $\mathcal{B} \subset \mathcal{K}$ if $\mathcal{B}$ is a general homogeneous space on Ω satisfying (90) and the a_j are continuous in Ω. First suppose that $a_j \geq \varepsilon > 0$ for $2 \leq j \leq n$, and some $\varepsilon > 0$. If $\delta(x,y)$ denotes the subunit distance function, and if we fix $y \in \Omega$, $0 < r < \delta \, dist\,(y, \partial\Omega)$, and set $f(x) = \delta(x,y)$, then (4) shows that f is Lipschitz continuous since

$$|f(x) - f(z)| \leq \delta(x,z) \leq \frac{|x-z|}{\varepsilon},$$

where the final inequality follows upon considering the subunit curve

$$\gamma(t) = x + \frac{\varepsilon}{|z-x|}t(z-x), \qquad 0 \leq t \leq \frac{|x-z|}{\varepsilon},$$

joining x to z. From the weaker version of (90), with (92) added to the right-hand side, we obtain

$$(95) \qquad |\delta(x,y) - C_{y,r}| \leq C \int_{B(y,Cr)} |\nabla_a f(z)| \frac{d(x,z)}{|B(x,d(x,z))|} dz$$

for $x \in B(y,r)$.

We now claim that

$$(96) \qquad |\nabla_a f(x)| \leq \sqrt{n} \qquad a.e. \; x, \qquad \text{for } f(x) = \delta(x,y).$$

This was proved in [12] and [11] with a larger constant and in the sense of distributions. Here we give an alternative simpler proof. Indeed, for $1 \leq j \leq n$ and $\beta < 1$, the curve $\gamma_j(t) = x + \beta t a_j(x) \mathbf{e}_j$ is subunit for t sufficiently small since $\gamma_j'(t) = \beta a_j(x) \mathbf{e}_j$ and $|\beta a_j(x)| < a_j(x + \beta t a_j(x) \mathbf{e}_j)$ for small t by the continuity of a_j. Thus $\delta(x, x + \beta t a_j(x) \mathbf{e}_j) \leq |t|$ for t small, and since δ is a metric,

$$\left| \beta a_j(x) \frac{\partial f}{\partial x_j}(x) \right| = \left| \lim_{t \to 0} \frac{f(x) - f(x + \beta t a_j(x) \mathbf{e}_j)}{t} \right|$$

$$\leq \limsup_{t \to 0} \left| \frac{\delta(x, x + \beta t a_j(x) \mathbf{e}_j)}{t} \right| \leq 1,$$

for $\beta < 1$ and $1 \leq j \leq n$. Thus $|\nabla_a f(x)| = \left(\sum_{j=1}^{n} \left| a_j(x) \frac{\partial f}{\partial x_j}(x) \right|^2 \right)^{\frac{1}{2}} \leq \sqrt{n}$.

It now follows from (95) and the inequality

$$(97) \qquad \int_{B(y,Cr)} \frac{d(x,z)}{|B(x,d(x,z))|} dz \leq C'r, \qquad x \in B(y,r),$$

that

$$(98) \qquad |\delta(x,y) - C_{y,r}| \leq Cr$$

for $x \in B(y, r)$. To see (97), we note that $B(y, Cr) \subset B(x, \gamma Cr)$ by the engulfing property of the balls, and so we compute

$$\int_{B(y,Cr)} \frac{d(x,z)}{|B(x,d(x,z))|} dz \leq \int_{B(x,\gamma Cr)} \frac{d(x,z)}{|B(x,d(x,z))|} dz$$

$$= \sum_{j=0}^{\infty} \int_{B(x,2^{-j}\gamma Cr) \backslash B(x,2^{-j-1}\gamma Cr)} \frac{d(x,z)}{|B(x,d(x,z))|} dz$$

$$\leq \sum_{j=0}^{\infty} \frac{2^{-j}\gamma Cr}{|B(x,2^{-j-1}\gamma Cr)|} \left| B(x,2^{-j}\gamma Cr) \backslash B(x,2^{-j-1}\gamma Cr) \right|$$

$$\leq \gamma C^2 \sum_{j=0}^{\infty} 2^{-j} r = C' r,$$

by the doubling property $|B(x, 2^{-j}\gamma Cr)| \leq C |B(x, 2^{-j-1}\gamma Cr)|$. Continuing with (98), we thus obtain

$$\delta(x,y) = \delta(x,y) - \delta(y,y) \leq |\delta(x,y) - C_{y,r}| + |\delta(y,y) - C_{y,r}| \leq Cr$$

for $x \in B(y, r)$, and so $B(y, r) \subset K(y, Ct)$ for $0 < r < t$ with a constant C independent of $0 < \varepsilon < 1$.

It remains to prove the general case. Given a_j, define $a_j^{\varepsilon} = a_j + \varepsilon$ for $0 < \varepsilon < 1$. Then (90) remains valid uniformly in $0 < \varepsilon < 1$ for a_j^{ε} in place of a_j, since neither the balls B nor the quasimetric d vary with ε. If we denote the corresponding subunit balls by $\mathcal{K}^{\varepsilon} = \{K^{\varepsilon}(y, r) : y \in \Omega, 0 < r < r_0\}$, then we have proven above that $\mathcal{B} \subset \mathcal{K}^{\varepsilon}$, i.e. $B(y, r) \subset K^{\varepsilon}(y, Cr)$, uniformly in $\varepsilon > 0$. We now show that

$$\cap_{0<\varepsilon<1} K^{\varepsilon}(y, r) \subset \cap_{0<\varepsilon<1} K\left(y, \sqrt{n}r + \varepsilon\right).$$

To prove this, let $x \in \cap_{0<\varepsilon<1} K^{\varepsilon}(y, r)$. Then for every $0 < \varepsilon < 1$, there is a Lipschitz curve $\gamma^{\varepsilon}(t) = \left(\gamma_j^{\varepsilon}(t)\right)_{j=1}^{n}$ satisfying (note that we may stop the curve γ^{ε} as soon as it hits x)

$$\gamma^{\varepsilon}(0) = y,$$
$$\gamma^{\varepsilon}(r) = x,$$
$$\left| \left(\gamma_j^{\varepsilon}\right)'(t) \right| \leq a_j\left(\gamma^{\varepsilon}(t)\right) + \varepsilon.$$

Since the a_j are bounded, the family $\{\gamma^{\varepsilon}(t)\}_{0<\varepsilon<1}$ is equicontinuous, and there is a continuous curve $\gamma(t)$ and a sequence $\{\varepsilon_i\}_{i=1}^{\infty}$ with $\lim_{i\to\infty} \varepsilon_i = 0$, such that $\lim_{i\to\infty} \gamma^{\varepsilon_i}(t) = \gamma(t)$ uniformly for $t \in [0, r]$. It follows easily that $\gamma(t)$ is Lipschitz and satisfies

$$\gamma(0) = y,$$
$$\gamma(r) = x,$$
$$|\gamma_j'(t)| \leq a_j(\gamma(t)),$$

where the third line follows by considering a fixed difference quotient and letting $i \to \infty$. Indeed,

$$
\begin{aligned}
\left| \frac{\gamma_j (t+h) - \gamma_j (t)}{h} \right| &= \lim_{i \to \infty} \left| \frac{\gamma_j^{\varepsilon_i} (t+h) - \gamma_j^{\varepsilon_i} (t)}{h} \right| \\
&\leq \liminf_{i \to \infty} \left\{ a_j \left(\gamma^{\varepsilon_i} (t + c_i h) \right) + \varepsilon_i \right\} \\
&= a_j \left(\gamma (t + ch) \right)
\end{aligned}
$$

for some $0 \leq c \leq 1$ upon taking a further subsequence such that $c_i \to c$, and using the uniform convergence of γ^{ε_i} to γ along with the continuity of a_j. Now let $h \to 0$ and use the continuity of a_j again to obtain $\left| \gamma_j' (t) \right| \leq a_j \left(\gamma (t) \right)$. Thus $\delta (x, y) \leq r\sqrt{n} < r\sqrt{n} + \varepsilon$ for all $\varepsilon > 0$, and we're done. Observe that the sharper containment

$$
\cap_{0 < \varepsilon < 1} K^{\varepsilon} (y, r) \subset \cap_{0 < \varepsilon < 1} K (y, r + \varepsilon)
$$

can be obtained by considering $\left(\sum_{j=1}^{n} \gamma_j' (t) \xi_j \right)^2$ in place of $\left| \gamma_j' (t) \right|^2$, as is done in the proof of Lemma 66 below.

REMARK 46. *The proof of Proposition 44 actually yields a bit more. To describe this, we introduce the family of rectangles $\mathcal{K}^* = \{ K^* (x, r), x \in \Omega, 0 < r < r_0 \}$, not necessarily arising from a homogeneous space, that are related to the subunit balls $\mathcal{K} = \{ K (x, r) \}$ as follows. Define*

$$
K_j^* (x, r) = \sup \left\{ \int_0^r a_j (\gamma (s)) \, ds : \gamma (s) \text{ is subunit}, \ \gamma (0) = x \right\},
$$

$$
K^* (x, r) = \prod_{j=1}^{n} \left[x_j - K_j^* (x, r), x_j + K_j^* (x, r) \right],
$$

for $x \in \Omega$, $0 < r < \delta \, \mathrm{dist} (x, \partial\Omega)$, so that by (94) we have

$$
K (x, r) \subset \widetilde{K} (x, r) \subset K^* (x, r), \qquad x \in \Omega, 0 < r < \delta \, \mathrm{dist} (x, \partial\Omega).
$$

The first half of the proof of Proposition 44 shows that if the size condition (91) holds, along with the other hypotheses on a_j, then $\mathcal{K}^* \subset \widetilde{\mathcal{B}}$.

REMARK 47. *Proposition 44 admits a more general extension to continuous quadratic forms $\mathcal{Q}$. Let $\mathcal{K}$ be related to $\mathcal{Q}$ as in Definition 4.*

(1) *Then $\mathcal{B} \subset \mathcal{K}$ if $\mathcal{B}$ is a general homogeneous space of balls $B (x, r)$ with quasimetric $d (x, y)$, and $\mathcal{Q} (x, \xi)$ is a continuous nonnegative semidefinite quadratic form satisfying the weak subrepresentation property that for each $y \in \Omega$, $0 < r < \delta \, \mathrm{dist} (y, \partial\Omega)$, there is a constant $C_{y,r}$ such that*

$$
|f (x) - C_{y,r}| \leq C \int_{B(y, C_0 r)} \| \nabla f (z) \|_{\mathcal{Q}} \frac{d (x, z)}{|B (x, d (x, z))|} dz, \qquad x \in B (y, r),
$$

for all f Lipschitz on $B (y, r)$.

(2) *Conversely, let $\mathcal{Q} (x, \xi) = \xi' Q (x) \xi = \sum_{i,j=1}^{n} q_{ij} (x) \xi_i \xi_j$ with $q_{11} (x) \equiv 1$ and set $A_j (x, r) = \int_0^r q_{jj} (x_1 + t, x_2, ..., x_n) \, dt$, $1 \leq j \leq n$. If the diagonal entries q_{jj} are Lipschitz continuous in x, and reverse Hölder of infinite order in x_i for $i \neq j$ uniformly in the remaining variables, and if $\mathcal{B}$ is a family of sets satisfying the size limiting condition (91), then $\widetilde{\mathcal{K}} \subset \widetilde{\mathcal{B}}$.*

The first assertion in the remark is proved as in Proposition 44, but using eigenvectors and eigenvalues of $Q(x)$ as in the proof of Proposition 68. As for the second assertion, the proof in Proposition 44 shows that $\widetilde{\mathcal{K}}_{diag} \subset \widetilde{\mathcal{B}}$ where $\mathcal{K}_{diag}$ denotes the subunit balls corresponding to the diagonal form $\mathcal{Q}_{diag}(x,\xi) = \sum_{j=1}^{n} q_{jj}(x)\xi_j^2$. Now $\widetilde{\mathcal{K}} \subset \widetilde{\mathcal{K}}_{diag}$ follows from the inequality $\mathcal{Q}(x,\xi) \le n\mathcal{Q}_{diag}(x,\xi)$:

$$\xi'Q\xi = \sum_{i,j=1}^{n} (\xi_i \mathbf{e}_i)' \, Q\,(\xi_j \mathbf{e}_j) \le \sum_{i,j=1}^{n} \frac{1}{2}\left\{ (\xi_i \mathbf{e}_i)' \, Q\,(\xi_i \mathbf{e}_i) + (\xi_j \mathbf{e}_j)' \, Q\,(\xi_j \mathbf{e}_j) \right\} = n\xi'Q_{diag}\xi.$$

3.1. Examples. We begin this subsection with a pathological example that graphically illustrates the breakdown of Sobolev and Poincaré inequalities for the subunit balls when continuity of the vector fields is violated.

EXAMPLE 48. *Let $n = 2$ and set $a_1 \equiv 1$ and $a_2(x_1,x_2) = \chi_{\mathbb{Q}}(x_1)$, where $\mathbb{Q}$ denotes the rational numbers. Then the flag balls degenerate to horizontal line segments, the divergence form operator $L = \frac{\partial}{\partial x_1} a_1 \frac{\partial}{\partial x_1} + \frac{\partial}{\partial x_2} a_2 \frac{\partial}{\partial x_2} = \frac{\partial^2}{\partial x_1^2}$ fails spectacularly to be subelliptic, and yet the subunit balls are equivalent to Euclidean balls. Indeed, the curves*

$$t \to (x_1 + t, x_2)$$

are subunit for all $x \in \mathbb{R} \times \mathbb{R}$, and the curves

$$t \to (x_1, x_2 + t)$$

are subunit for all $x \in \mathbb{Q} \times \mathbb{R}$. Any two points x and y can thus be joined by a subunit curve of length $|x_1 - y_1| + |x_2 - y_2|$, by proceeding horizontally from (x_1, x_2) to (q, x_2), where q is rational and lies between x_1 and y_1, then vertically from (q, x_2) to (q, y_2), and finally horizontally from (q, y_2) to (y_1, y_2).

The difficulty in the above example is that the vertical subunit curves cannot be perturbed in the horizontal direction, since a_2 fails to be continuous anywhere. This precludes the possibility of a subrepresentation inequality for $f(x)$ with absolutely continuous kernel.

The following example of analytic vector fields in $\mathbb{R}^3$ satisfies the flag condition with $m = 2$ and $\#\mathcal{I}_1 = 1$, but the rectangular boxes $A(x,r)$ fail to satisfy the engulfing property, and hence fail to satisfy Definition 18 as well.

EXAMPLE 49. *Let*

$$a_1(x) = 1,$$
$$a_2(x) = 3x_1^2,$$
$$a_3(x) = 9x_1^8 + x_2^2.$$

With $y = (0, y_2, 0)$, $0 \le y_2 \le 1$, we have

$$A_2(y,r) = \int_0^r a_2(t, y_2, 0)\, dt = r^3,$$

$$A_3(y,r) = \int_0^r a_3(t, y_2, 0)\, dt = r^9 + y_2^2 r.$$

Then the rectangular boxes $A((0,0,0),r)$ and $A\left((0,r^3,0),r^2\right)$ are given by

$$A((0,0,0),r) = [-r,r] \times \left[-r^3, r^3\right] \times \left[-r^9, r^9\right],$$

$$A\left((0,r^3,0),r^2\right) = \left[-r^2, r^2\right] \times \left[r^3 - r^6, r^3 + r^6\right] \times \left[-r^8 - r^{18}, r^8 + r^{18}\right],$$

and neither is contained in a fixed multiple of the other, uniformly in $0 < r < 1$, despite having nonempty intersection. On the other hand, one easily computes that for $0 \leq y_2 \leq r$, the flag balls $B\left((0, y_2, 0), r\right)$, which do *satisfy the engulfing property, are comparable to the rectangular boxes*

$$\left[-r, r\right] \times \left[y_2 - r^3, y_2 + r^3\right] \times \left[-y_2^2 r - r^7, y_2^2 r + r^7\right].$$

CHAPTER 3

Proof of the general subellipticity theorem

The purpose of this section is to prove Theorem 8 by the Moser iteration method. We begin by establishing in subsection 3.1 a reverse Sobolev inequality of Caccioppoli type for weak (sub, super) solutions u to (22). Building on an idea of Taylor [44], our approach is to compute the equation satisfied by u^β, $\beta \in \mathbb{R}$, and then estimate its energy. Then in subsection 3.2 we will use our homogeneous space structure with doubling (13), the Sobolev inequality (15), and the "accumulating sequence of Lipschitz cutoff functions" condition (20) to iterate the reverse Sobolev inequality against (15) to obtain local boundedness of weak solutions. In subsection 3.3 we invoke the Poincaré inequality (17) and the John-Nirenberg theorem on homogeneous spaces to obtain a strong Harnack inequality for nonnegative weak solutions. Finally, following Moser, who followed De Giorgi, we iterate the Harnack inequality in subsection 3.4, and use the containment condition (11), to obtain Hölder continuity of weak solutions to (22).

PROOF (OF THEOREM 8). Let $L = \nabla' B(x) \nabla$ where $B(x)$ satisfies (27), and consider the linear operator

$$\mathcal{L} \equiv L + \mathbf{H}\mathbf{R} + \mathbf{S}'\mathbf{G} + F$$

where $\mathbf{R} = \{R_i\}_{i=1}^N$ and $\mathbf{S} = \{S_i\}_{i=1}^N$ are collections of vector fields subunit with respect to $B(x)$, and F, $\mathbf{G} = \{G_i\}_{i=1}^N$ and $\mathbf{H} = \{H_i\}_{i=1}^N$ are measurable functions. We must show that there is

$$\alpha = \alpha(K, \varepsilon, p, q, c_{sym}, C_{sym}, N_q) > 0$$

such that every weak solution u of the equation

(99) $$\mathcal{L}u = f + \mathbf{T}'\mathbf{g}$$

in Ω satisfies

(100) $$\|u\|_{C^\alpha(K)} \leq C\left(K, p, q, c_{sym}, C_{sym}, N_q, N_q', \|u\|_{L^2(\Omega)}\right)$$

for all compact subsets K of Ω, inhomogeneous data f and $\mathbf{g} = \{g_i\}_{i=1}^N$ satisfying (24) and operator coefficients F, $\mathbf{G} = \{G_i\}_{i=1}^N$ and $\mathbf{H} = \{H_i\}_{i=1}^N$ satisfying (23) for some $q > Q = \max\{Q^*, 2\sigma'\}$, and collections of subunit vector fields $\mathbf{R} = \{R_i\}_{i=1}^N$, $\mathbf{S} = \{S_i\}_{i=1}^N$ and $\mathbf{T} = \{T_i\}_{i=1}^N$. Here ε is as in (11), $p > \max\{2\sigma', 4\}$ is as in (20), σ is as in (15) and Q^* is as in (9). Recall the definition of a classical weak (sub, super) solution:

DEFINITION 50. *A function $u \in W^{1,2}(\Omega)$ is a weak* $\begin{pmatrix} solution \\ subsolution \\ supersolution \end{pmatrix}$ *of*

(101)
$$Lu + \mathbf{H R}u + \mathbf{S}'\mathbf{G}u + Fu = f + \mathbf{T}'\mathbf{g}$$

in Ω if
(102)
$$-\int (\nabla u)' \, B\nabla w + \int (\mathbf{H R}u)\, w + \int u\mathbf{G S}w + \int Fuw \begin{pmatrix} = \\ \geq \\ \leq \end{pmatrix} \int fw + \int \mathbf{g T}w,$$

for all nonnegative $w \in W_0^{1,2}(\Omega)$.

In the case of a weak solution, we may equivalently test (102) over all $w \in W_0^{1,2}(\Omega)$. Here the juxtaposition of vectors in $\mathbf{H R}$, $\mathbf{G S}$ and $\mathbf{g T}$ means $\sum_{i=1}^{N} H_i R_i$, $\sum_{i=1}^{N} G_i S_i$ and $\sum_{i=1}^{N} g_i T_i$ respectively. Note that the integrals in (102) converge absolutely since $u \in W^{1,2}(\Omega)$, $w \in W_0^{1,2}(\Omega)$, together with the Sobolev inequality (15), imply that

(103)
$$u \in L_{loc}^{2\sigma}(\Omega)\,, w \in L^{2\sigma}(\Omega)\,,$$

and so by Hölder's inequality,
$$uw \in L^{\sigma}(\Omega)\,,$$

while our assumptions on the coefficients and data imply

(104)
$$\begin{aligned} F, f &\in L^{\sigma'}(\Omega)\,, \\ \mathbf{G}, \mathbf{H}, \mathbf{g} &\in L^{2\sigma'}(\Omega)\,, \end{aligned}$$

which by Hölder's inequality yields

(105)
$$\mathbf{G}u, \mathbf{H}w \in L^2(\Omega)\,.$$

Throughout this proof, we often use the notation
$$\begin{aligned} \langle U, W \rangle &= U'BW, \\ \|U\| &= \langle U, U \rangle^{\frac{1}{2}}\,, \end{aligned}$$

where the underlying matrix is always assumed to be B as in the operator L under question. Note by (27) that $\|U\|^2 \approx \mathcal{Q}(x, U) = \|U\|_{\mathcal{Q}}$ where $\|\cdot\|_{\mathcal{Q}}$ is given in (14).

REMARK 51. *For the notion of weak solution we are using here, namely that of a solution $u \in W^{1,2}(\Omega)$ with test functions $w \in W_0^{1,2}(\Omega)$, we can significantly relax the conditions on the data in order to make sense of (102). In fact, the usual Sobolev embedding theorem in Euclidean space shows that $u, w \in W^{1,2}(\Omega) \subset L_{loc}^{\frac{2n}{n-2}}(\Omega)$ for $n \geq 3$, and so we need only assume $f, F \in L_{loc}^{\frac{q}{2}}(\Omega)$ and $\mathbf{g}, \mathbf{G}, \mathbf{H} \in L_{loc}^{q}(\Omega)$ for $q = n$ (and $q > 2$ in the case $n = 2$). Note that since the balls $B(x, r)$ in our homogeneous space are by hypothesis contained in the Euclidean balls $D(x, \frac{r}{c})$, we have*
$$c_n r^n = \left| D\left(x, \frac{r}{c}\right) \right| \geq |B(x, r)| \geq c_q r^q$$

for any $q > Q$, and it follows that $Q \geq n$. In the appendix, we will consider other notions of weak solution to (99), in particular that of $u \in W_{\mathcal{Q}}^{1,2}(\Omega)$ with test functions in $\left(W_{\mathcal{Q}}^{1,2}\right)_0 (\Omega)$. Here $W_{\mathcal{Q}}^{1,2} (\Omega)$ denotes the completion of $Lip_1 (\Omega)$ under the norm

$$\|w\|_{W_{\mathcal{Q}}^{1,2}(\Omega)} = \left\{ \int_\Omega \left(|w|^2 + \|\nabla w\|_{\mathcal{Q}}^2 \right) \right\}^{\frac{1}{2}}.$$

The integrals in (102) also make sense in this setting by the argument given above using (15), (23) and (24). However, the gradients of elements of $W_{\mathcal{Q}}^{1,2}$ are no longer functions, but rather certain Cauchy sequences. Further details are in subsection 6.7 of the appendix.

1. $\mathcal{W}$-weak solutions and admissible compositions

We will employ an equality (see (117) below) which shows that certain nonlinear operations on (sub, super) solutions to an equation yield (sub, super) solutions to a related equation, but with a weaker notion of (sub, super) solution than the classical one in Definition 50. We begin by introducing this weaker notion of weak (sub, super) solution. We emphasize that this weaker definition, and the variant that follows, is used only in the course of implementing Moser iteration in this proof, and does not appear in statements of any theorems or propositions.

DEFINITION 52. *Let $\mathcal{W}$ be a subset of the nonnegative elements in $W_0^{1,2} (\Omega)$. We say that a function $u \in W^{1,2} (\Omega)$ is a $\mathcal{W}$-weak $\begin{pmatrix} \text{solution} \\ \text{subsolution} \\ \text{supersolution} \end{pmatrix}$ of the divergence form equation (101) if the the integrals in (102) are absolutely convergent and the indicated (in)equality holds for all $w \in \mathcal{W}$.*

We also need a corresponding notion of $\mathcal{W}$-weak sense for more general equations (inequalities) of the form

$$(106) \qquad \sum_{j=1}^N \mathcal{H}_j \left(\mathcal{T}_j' \mathcal{G}_j \right) \begin{pmatrix} = \\ \geq \\ \leq \end{pmatrix} \mathcal{F},$$

where $\mathcal{F}$ is a function, $\mathcal{G} = (\mathcal{G}_1, ..., \mathcal{G}_N)$ and $\mathcal{H} = (\mathcal{H}_1, ..., \mathcal{H}_N)$ are collections of functions, and $\mathcal{T}' = (\mathcal{T}_1', ..., \mathcal{T}_N')$ is a collection of transposed subunit vector fields (with respect to the matrix $B(x)$). The precise properties imposed on these functions and vector fields will be described in Definition 53 below. The relation (106) includes equations such as (116) below. Indeed, as we will see, $\mathcal{F}$ can include expressions of the form $h'' (u) (\mathbf{T}u) \mathbf{g} + h'' (u) \|\nabla u\|^2$, while $\sum_{j=1}^N \mathcal{H}_j \mathcal{T}_j' \mathcal{G}_j$ can include expressions of the form $\mathbf{T}' (h' (u) \mathbf{g}) - \mathbf{S}' (h' (u) u\mathbf{G})$ with $\mathcal{H}_j \equiv 1$, as well as $h' (u) Lu$. For the latter, let $N = n$, $\mathcal{H}_j = h' (u)$, $\mathcal{T}_j = \frac{\partial}{\partial x_j}$ and $\mathcal{G}_j = [B(x) \nabla u]_j$, the jth component of the vector $B(x) \nabla u$. The classical meaning attached to the relation (106) in the weak sense is that

$$(107) \qquad \sum_{j=1}^N \int_\Omega \mathcal{T}_j (w\mathcal{H}_j) \mathcal{G}_j \begin{pmatrix} = \\ \geq \\ \leq \end{pmatrix} \int_\Omega w\mathcal{F}$$

holds for all nonnegative $w \in W_0^{1,2} (\Omega)$, provided the functions $\mathcal{F}, \mathcal{G}, \mathcal{H}$ and vector fields $\mathcal{T}$ in (106) result in absolutely convergent integrals in (107). We generalize this as follows.

DEFINITION 53. *Let $\mathcal{W}$ be a subset of the nonnegative elements in $W_0^{1,2}(\Omega)$. Then we say that (106) holds in the $\mathcal{W}$-weak sense if the integrals in (107) are absolutely convergent and the indicated (in)equality holds for all $w \in \mathcal{W}$.*

Note that Definition 53 generalizes Definition 52 in the sense that $u \in W^{1,2}(\Omega)$

is a $\mathcal{W}$-weak $\begin{pmatrix} \text{solution} \\ \text{subsolution} \\ \text{supersolution} \end{pmatrix}$ of the divergence form equation (101) if and only

if (101) holds in the $\mathcal{W}$-weak sense. The point of introducing the notion of $\mathcal{W}$-weak sense for the more general equations (inequalities) (106), is that the equations satisfied by nonlinear functions of solutions u to (101), such as (111) below, are no longer of the standard divergence form given in (101).

Now let $u \in W^{1,2}(\Omega)$. We will compose u with nonlinear functions h of the following "admissible" form.

DEFINITION 54. *Let I be an interval and $h \in C^1(I) \cap C^2_{pw}(I)$ be positive and monotone (i.e. either nondecreasing on I or nonincreasing on I), where $C^2_{pw}(I)$ is the space of piecewise twice continuously differentiable functions on I. The function h is said to be* admissible *on I if there is a positive constant C such that*

$$(108) \qquad |h'(t)|, |h''(t)|, |th''(t)| \leq C, \quad t \in I.$$

Moreover, given $u \in W^{1,2}(\Omega)$, we say that h is admissible for u *if h is admissible on some interval I containing the range of u.*

We emphasize that the constant C in (108) is qualitative only, and does not appear in any estimates for solutions to equations. If h is admissible on I, then in particular h satisfies

$$(109) \qquad \begin{aligned} 0 \quad &< \quad h(t) \leq C(1+|t|), \\ |h(t)h''(t)| \quad &\leq \quad C(|h''(t)| + |th''(t)|) \leq C', \end{aligned}$$

for $t \in I$. If h is admissible for u, it follows that $\widetilde{u}(x) = h(u(x))$ lies in $W^{1,2}(\Omega)$ since Ω is bounded (see e.g. Lemma 7.5 in [14]).

In the sequel we shall use three different subsets $\mathcal{W}$ of the nonnegative elements in $W_0^{1,2}(\Omega)$. These are conveniently summarized in Remark 57 below, and we define the first of these now. We claim that if h is admissible for u and

$$(110) \qquad \mathcal{M}[u; h] = \left\{ w \in W_0^{1,2}(\Omega) : w \geq 0 \text{ and } h'(u)w \in W_0^{1,2}(\Omega) \right\},$$

then

$$(111) \qquad L\widetilde{u} = h'(u)Lu + h''(u)\|\nabla u\|^2$$

holds in the $\mathcal{M}[u; h]$-weak sense. According to Definition 53, this means that

$$(112) \qquad -\int [\nabla w]' \, B\nabla\widetilde{u} = -\int [\nabla(wh'(u))]' \, B\nabla u + \int wh''(u)\|\nabla u\|^2,$$

and where the integrals on the right are absolutely convergent, for all $w \in \mathcal{M}[u; h]$. First note that all three integrals in (112) are absolutely convergent since $\nabla u \in L^2(\Omega)$ and $w, wh'(u) \in W_0^{1,2}(\Omega)$, which implies $wh''(u)\nabla u \in L^2(\Omega)$: the product rule (see e.g. (7.18) in [14]) yields

$$(113) \qquad wh''(u)\nabla u = \nabla(wh'(u)) - (\nabla w)h'(u) \in L^2(\Omega).$$

The integral equality (112) follows from the product rule and chain rule upon noting the absolute convergence of the integrals involved:

$$\int \left[\nabla \left(wh'\left(u\right)\right)\right]' B\nabla u \;\; = \;\; \int h'\left(u\right)\left[\nabla w\right]' B\nabla u + \int w\left[\nabla \left(h'\left(u\right)\right)\right]' B\nabla u$$

$$= \;\; \int \left[\nabla w\right]' B\nabla h\left(u\right) + \int wh''\left(u\right)\left[\nabla u\right]' B\nabla u$$

$$= \;\; \int \left[\nabla w\right]' B\nabla \widetilde{u} + \int wh''\left(u\right) \|\nabla u\|^{2}.$$

We will also need the two equalities (recall that juxtaposition of vectors implies the dot product)

(114)
$$h'\left(u\right)\mathbf{T}'\mathbf{g} \;\; = \;\; \mathbf{T}'\left(h'\left(u\right)\mathbf{g}\right) + h''\left(u\right)\left(\mathbf{T}u\right)\mathbf{g}$$
$$h'\left(u\right)\mathbf{S}'\mathbf{G}u \;\; = \;\; \mathbf{S}'\left(h'\left(u\right)\mathbf{G}u\right) + h''\left(u\right)\left(\mathbf{S}u\right)\mathbf{G}u$$

in the usual weak sense for $\mathbf{G}, \mathbf{g} \in L^{n}\left(\Omega\right)$: if we write $\mathbf{U} = \mathbf{T}$ and $\mathbf{h} = \mathbf{g}$ for the first equality, and $\mathbf{U} = \mathbf{S}$ and $\mathbf{h} = \mathbf{G}u$ for the second, then the weak sense has the meaning

$$\int w\left(h'\left(u\right)\mathbf{U}'\mathbf{h}\right) \;\; = \;\; \int \mathbf{U}wh'\left(u\right) \cdot \mathbf{h}$$

$$= \;\; \int h'\left(u\right)\mathbf{U}w \cdot \mathbf{h} + \int wh''\left(u\right)\mathbf{U}u \cdot \mathbf{h}$$

$$= \;\; \int w\mathbf{U}'\left(h'\left(u\right)\mathbf{h}\right) + \int wh''\left(u\right)\left(\mathbf{U}u\right) \cdot \mathbf{h},$$

for all nonnegative $w \in W_{0}^{1,2}\left(\Omega\right)$. Note that the two integrals on the right-hand side of the middle line are absolutely convergent in both cases. Indeed, in the case $n \geq 3$, if $\mathbf{U} = \mathbf{T}$ and $\mathbf{h} = \mathbf{g}$, then $w \in W_{0}^{1,2}\left(\Omega\right) \subset L_{0}^{2\left(\frac{n}{2}\right)'}\left(\Omega\right)$ and $\mathbf{g} \in L^{2\left(\frac{n}{2}\right)}\left(\Omega\right)$ implies $w\mathbf{g} \in L^{2}\left(\Omega\right)$. Thus the second integral $\int wh''\left(u\right)\mathbf{T}u \cdot \mathbf{g}$ is absolutely convergent since $h''\left(u\right)$ is bounded and $\mathbf{T}u \in L^{2}\left(\Omega\right)$. If $\mathbf{U} = \mathbf{S}$ and $\mathbf{h} = \mathbf{G}u$, then $w\mathbf{G} \in L^{2}\left(\Omega\right)$ as above, and the second integral $\int wh''\left(u\right)\mathbf{S}u \cdot \mathbf{G}u$ is absolutely convergent since $uh''\left(u\right)$ is bounded and $\mathbf{S}u \in L^{2}\left(\Omega\right)$. The first integral is also absolutely convergent in both cases. Similar comments apply when $n = 2$ using that $\mathbf{G}, \mathbf{g} \in L^{q}$ for some $q > 2$.

We are now ready to compute the equation satisfied by a nonlinear function of a weak solution to (99). Let $u \in W^{1,2}\left(\Omega\right)$, h be admissible for u and set $\widetilde{u} = h\left(u\right)$. Then from (111), we have in the $\mathcal{M}\left[u; h\right]$-weak sense,

(115)
$$L\widetilde{u} + \mathbf{H}\mathbf{R}\widetilde{u} = h'\left(u\right)\left(Lu + \mathbf{H}\mathbf{R}u\right) + h''\left(u\right)\|\nabla u\|^{2}.$$

Now assume in addition that u is a weak solution of (99) or (101) in Ω. Combining the above equalities (111), (114) and (115), we obtain that

(116)
$$L\widetilde{u} + \mathbf{H}\mathbf{R}\widetilde{u}$$
$$= \;\; h'\left(u\right)\left(f + \mathbf{T}'\mathbf{g} - \mathbf{S}'\mathbf{G}u - Fu\right) + h''\left(u\right)\|\nabla u\|^{2}$$
$$= \;\; h'\left(u\right)f + \mathbf{T}'\left(h'\left(u\right)\mathbf{g}\right) + h''\left(u\right)\left(\mathbf{T}u\right)\mathbf{g} + h''\left(u\right)\|\nabla u\|^{2}$$
$$- \mathbf{S}'\left(h'\left(u\right)u\mathbf{G}\right) - h''\left(u\right)\left(\mathbf{S}u\right)\mathbf{G}u - uh'\left(u\right)F$$

in the $\mathcal{M}\left[u; h\right]$-weak sense with $\mathcal{M}\left[u; h\right]$ as in (110). Indeed, the integral form of (116) follows using first (102) with w replaced by $wh'\left(u\right) \in W_{0}^{1,2}\left(\Omega\right)$ if $h' \geq 0$, otherwise use $-wh'(u)$, and then (112) to obtain

$$-\int wh'(u)\,f - \int \mathbf{T}\left(wh'(u)\right)\mathbf{g} + \int wh'(u)\,\mathbf{HR}u$$

$$+ \int \mathbf{S}\left(wh'(u)\right)\mathbf{G}u + \int wh'(u)\,Fu$$

$$= \int \left[\nabla\left(wh'(u)\right)\right]' B\nabla u$$

$$= \int \left[\nabla w\right]'\,B\nabla\widetilde{u} + \int wh''(u)\,\|\nabla u\|^2,$$

and finally applying (114) (which justifies the product rule) to the second and fourth integrals on the left. Thus if we define the operator

$$\widetilde{\mathcal{L}} \equiv L + \mathbf{HR} + \mathbf{S}'\widetilde{\mathbf{G}} + \widetilde{F}$$

where

$$\widetilde{F} = \frac{h'(u)\,u}{h(u)}F, \quad \widetilde{\mathbf{G}} = \frac{h'(u)\,u}{h(u)}\mathbf{G},$$

and let

$$\widetilde{f} = h'(u)\,f, \quad \widetilde{\mathbf{g}} = h'(u)\,\mathbf{g},$$

then in the sense of Definition 52, $\widetilde{u} = h(u)$ is a $\mathcal{M}[u;h]$-weak solution of the equation

(117) $$\widetilde{\mathcal{L}}\widetilde{u} = \widetilde{f} + \mathbf{T}'\widetilde{\mathbf{g}} + h''(u)\,\|\nabla u\|^2 + \Phi$$

in Ω, where

$$\Phi = h''(u)\left\{(\mathbf{T}u)\,\mathbf{g} - (\mathbf{S}u)\,\mathbf{G}u\right\},$$

and $\mathcal{M}[u;h]$ is as in (110). We remind the reader that for $w \in \mathcal{M}[u;h]$ the integral $\int w\Phi = \int wh''(u)\left\{(\mathbf{T}u)\,\mathbf{g} - (\mathbf{S}u)\,\mathbf{G}u\right\}$ is absolutely convergent since

$$\left|wh''(u)\,(\mathbf{T}u)\right|, \left|wh''(u)\,(\mathbf{S}u)\right| \leq \sqrt{N}\,\left|wh''(u)\right|\,\|\nabla u\| \leq C\left|wh''(u)\,\nabla u\right|$$

which is in $L^2(\Omega)$ by (113), and since $\mathbf{g}$ and $\mathbf{G}u$ are in $L^2(\Omega)$ by (104) and (105).

REMARK 55. *Suppose that $u \in W^{1,2}(\Omega)$ and h is admissible for u. If u is a weak $\left(\begin{smallmatrix}\text{subsolution}\\\text{supersolution}\end{smallmatrix}\right)$ of (101) in Ω and $h'(u)\left(\begin{smallmatrix}\geq\\\leq\end{smallmatrix}\right)0$, then the term $h'(u)\,(Lu + \mathbf{HR}u)$ in (115) satisfies*

$$h'(u)\,(Lu + \mathbf{HR}u) \geq h'(u)\,(f + \mathbf{T}'\mathbf{g} - \mathbf{S}'\mathbf{G}u - Fu)$$

in the $\mathcal{M}[u;h]$-weak sense, and so $\widetilde{u}$ is a $\mathcal{M}[u;h]$-weak subsolution of (117) in Ω, where $\mathcal{M}[u;h]$ is as in (110). Similarly, if u is a weak $\left(\begin{smallmatrix}\text{subsolution}\\\text{supersolution}\end{smallmatrix}\right)$ of (101) in Ω and $h'(u)\left(\begin{smallmatrix}\leq\\\geq\end{smallmatrix}\right)0$, then $\widetilde{u}$ is a $\mathcal{M}[u;h]$-weak supersolution of (117) in Ω. Note that the case of a weak solution of (101) is included here since h' has only one sign.

In the above, we have shown in particular that if u is a weak $\left(\begin{smallmatrix}\text{subsolution}\\\text{supersolution}\end{smallmatrix}\right)$ of (101), h is admissible for u and $h'(u) \geq 0$, then $\widetilde{u} = h(u)$ is a $\mathcal{M}[u;h]$-weak $\left(\begin{smallmatrix}\text{subsolution}\\\text{supersolution}\end{smallmatrix}\right)$ of (117). The subset $\mathcal{M}[u;h]$ is maximal for this purpose as is evident from the first equality in (116), where it is required that $w \in \mathcal{M}\,[u;h]$ satisfy

$$\int \left\{ wh'\,(u)\,(Lu + \mathbf{H}\mathbf{R}u) \right\} = \int \left\{ wh'\,(u)\,(f + \mathbf{T}'\mathbf{g} - \mathbf{S}'\mathbf{G}u - Fu) \right\}.$$

However, in applying Moser iteration to $\widetilde{u}$, we will see below that we do not need the full force of the conclusion that $\widetilde{u}$ is a $\mathcal{M}\,[u;h]$-weak $\left(\begin{array}{c} \text{subsolution} \\ \text{supersolution} \end{array} \right)$ of (117), but only that $\widetilde{u}$ is a $\mathcal{E}\,[\widetilde{u}]$-weak $\left(\begin{array}{c} \text{subsolution} \\ \text{supersolution} \end{array} \right)$ of (117) where the subset $\mathcal{E}\,[u]$ of "energy test functions" associated to a nonnegative u is given by

(118)
$$\mathcal{E}\,[u] = \left\{ \psi^2 u : \psi \in C_0^{0,1}\,(\Omega) \right\}.$$

To see that $\mathcal{E}\,[\widetilde{u}]$ is a subset of $\mathcal{M}\,[u;h]$, we must show $\psi^2 \widetilde{u} \in \mathcal{M}\,[u;h]$ for $\psi \in C_0^{0,1}\,(\Omega)$ and $\widetilde{u} = h\,(u)$, i.e.
(119)
$$\psi^2 h\,(u)\,,\psi^2 h\,(u)\,h'\,(u) \in W_0^{1,2}\,(\Omega)\,, \quad \text{if } h \text{ is admissible for } u \text{ and } \psi \in C_0^{0,1}\,(\Omega)\,.$$

We've already observed after (109) that $\psi^2 h\,(u) \in W_0^{1,2}\,(\Omega)$, and thus we have

$$\psi^2 h\,(u)\,h'\,(u) \in L^2\,(\Omega)$$

since $h'\,(u)$ is bounded. Moreover,

$$\left| \nabla \left(\psi^2 h\,(u)\,h'\,(u) \right) \right| \le \left| 2\psi h\,(u)\,h'\,(u)\,\nabla\psi \right| + \left| \psi^2 h'\,(u)^2\,\nabla u \right| + \left| \psi^2 h\,(u)\,h''\,(u)\,\nabla u \right| \in L^2\,(\Omega)$$

by (108) and (109). This completes the demonstration that $\mathcal{E}\,[\widetilde{u}] \subset \mathcal{M}\,[u;h]$ if h is admissible for u and $\widetilde{u} = h\,(u)$.

1.1. Energy sub and super solutions. Now let $u \in W^{1,2}\,(\Omega)$ and $\widetilde{u} = h\,(u)$, where h is admissible for u, and suppose that $\widetilde{u}$ is a positive $\mathcal{E}\,[\widetilde{u}]$-weak $\left(\begin{array}{c} \text{subsolution} \\ \text{supersolution} \end{array} \right)$ of (117) in Ω, which we sometimes refer to loosely by saying that $\widetilde{u}$ is an *energy* (sub or super) solution of (117) in Ω. Let $\psi \in C_0^{0,1}\,(\Omega)$. Then from the integral form of (117) we obtain

(120)
$$\int \left\langle \nabla h\,(u), \nabla \psi^2 h\,(u) \right\rangle + \int \psi^2 h\,(u)\,h''\,(u)\,\|\nabla u\|^2$$

$$\left(\begin{array}{c} \le \\ \ge \end{array} \right) - \int \widetilde{f}\psi^2 h\,(u) - \sum_{i=1}^{N} \int \widetilde{g}_i T_i\,(\psi^2 h\,(u)) + \sum_{i=1}^{N} \int (H_i R_i h\,(u))\,\psi^2 h\,(u)$$

$$- \sum_{i=1}^{N} \int h\,(u)\,\widetilde{G}_i S_i\,(\psi^2 h\,(u)) + \int \widetilde{F}h\,(u)\,\psi^2 h\,(u)$$

$$- \int \psi^2 h\,(u)\,h''\,(u)\,\{(\mathbf{T}u)\,\mathbf{g} - (\mathbf{S}u)\,\mathbf{G}u\}.$$

The left side of (120) equals

(121)
$$\int \psi^2 h'(u)^2 \langle \nabla u, \nabla u \rangle + 2 \int \psi h(u) h'(u) \langle \nabla u, \nabla \psi \rangle$$
$$+ \int \psi^2 h(u) h''(u) \|\nabla u\|^2$$
$$= \int \psi^2 \Gamma(u) \|\nabla u\|^2 + 2 \int \langle \psi \nabla h(u), h(u) \nabla \psi \rangle$$

where

$$\Gamma(t) = h'(t)^2 + h(t) h''(t) = \left(\frac{1}{2} h(t)^2 \right)''.$$

The first term $\mathcal{T} = \int \psi^2 \Gamma(u) \|\nabla u\|^2$ on the right-hand side of (121) will turn out to be the main term in (120) in our specific calculations below. Indeed, for the functions $h(t)$ that we consider, **either** $\Gamma(t) > 0$ everywhere (i.e. $h(t)^2$ is strictly convex), **or** $\Gamma(t) < 0$ everywhere (i.e. $h(t)^2$ is strictly concave), and in either case, all of the remaining terms in (120) and (121) have absolute value dominated by the sum of a small multiple of $|\mathcal{T}|$, a small mutiple of $\int \psi^2 h'(u)^2 \|\nabla u\|^2$, and a large multiple of integrals involving no derivatives of u. In preparation for proving this, we establish the following inequalities.

For $0 < \varepsilon < 1$ (this ε is not related to the ε in the containment condition (11)), we can estimate the last term on the right side above by

$$2 \left| \int \langle \psi \nabla h(u), h(u) \nabla \psi \rangle \right| \leq \varepsilon \int \langle \psi \nabla h(u), \psi \nabla h(u) \rangle + \varepsilon^{-1} \int \langle h(u) \nabla \psi, h(u) \nabla \psi \rangle$$
$$= \varepsilon \int \psi^2 h'(u)^2 \|\nabla u\|^2 + \varepsilon^{-1} \int h(u)^2 \|\nabla \psi\|^2.$$

The three terms involving a sum $\sum_{i=1}^{N}$ on the right side of (120) can be estimated by dominating

$$\left| \int \widetilde{g}_i T_i \left(\psi^2 h(u) \right) \right| = \left| 2 \int \psi h(u) h'(u) g_i (T_i \psi) + \int \psi^2 h'(u)^2 g_i (T_i u) \right|$$

by

$$\int h(u)^2 (T_i \psi)^2 + \int \psi^2 h'(u)^2 g_i^2$$
$$+ \varepsilon \int \psi^2 h'(u)^2 (T_i u)^2 + \varepsilon^{-1} \int \psi^2 h'(u)^2 g_i^2$$
$$\leq \int h(u)^2 \|\nabla \psi\|^2 + \int \psi^2 h'(u)^2 g_i^2$$
$$+ \varepsilon \int \psi^2 h'(u)^2 \|\nabla u\|^2 + \varepsilon^{-1} \int \psi^2 h'(u)^2 g_i^2,$$

since $(T_i u)^2 \leq \|\nabla u\|^2$ and $(T_i \psi)^2 \leq \|\nabla \psi\|^2$ follow from the hypothesis that T_i is subunit; similarly,

$$\left| \int h(u) \widetilde{G}_i S_i \left(\psi^2 h(u) \right) \right|$$

$$\leq \int h(u)^2 \|\nabla \psi\|^2 + \int \psi^2 h'(u)^2 u^2 G_i^2$$

$$+\varepsilon \int \psi^2 h'(u)^2 \|\nabla u\|^2 + \varepsilon^{-1} \int \psi^2 h'(u)^2 u^2 G_i^2,$$

since S_i is subunit; and finally

$$\left| \int (H_i R_i h(u)) \psi^2 h(u) \right| \leq \varepsilon \int \psi^2 h'(u)^2 \|\nabla u\|^2 + \varepsilon^{-1} \int \psi^2 h(u)^2 H_i^2,$$

since R_i is subunit.

We now consider the final term on the right-hand side of (120). Here we use $|hh''| \leq |\Gamma| + |h'|^2$ to obtain

$$\left| \int \psi^2 h(u) h''(u) \left\{ (\mathbf{T}u) \, \mathbf{g} - (\mathbf{S}u) \, \mathbf{G}u \right\} \right| \leq \varepsilon \int \psi^2 |h(u) h''(u)| \left\{ |\mathbf{T}u|^2 + |\mathbf{S}u|^2 \right\}$$

$$+ \varepsilon^{-1} \int \psi^2 |h(u) h''(u)| \left\{ |\mathbf{g}|^2 + |\mathbf{G}|^2 u^2 \right\}$$

$$\leq 2N\varepsilon \int \psi^2 \left(|\Gamma(u)| + |h'(u)|^2 \right) \|\nabla u\|^2$$

$$+ \varepsilon^{-1} \int \psi^2 |h(u) h''(u)| \left\{ |\mathbf{g}|^2 + |\mathbf{G}|^2 u^2 \right\}.$$

Now let $u \in W^{1,2}(\Omega)$ and $\widetilde{u} = h(u)$ where h is admissible for u, and assume that $\widetilde{u}$ satisfies at least one of the following:

(i) $\Gamma(t) > 0$ on I and $\widetilde{u}$ is a $\mathcal{E}[\widetilde{u}]$-weak subsolution of (117) in Ω,
(ii) $\Gamma(t) < 0$ on I and $\widetilde{u}$ is a $\mathcal{E}[\widetilde{u}]$-weak supersolution of (117) in Ω.

Then for $0 < \varepsilon < 1$, we obtain by combining the above estimates that

$$(122) \qquad \int \psi^2 \left[(1 - 2N\varepsilon) |\Gamma(u)| - (5N+1)\varepsilon h'(u)^2 \right] \|\nabla u\|^2$$

$$\leq \left(2N + \varepsilon^{-1} \right) \int h(u)^2 \|\nabla \psi\|^2 + \varepsilon^{-1} \int \psi^2 h(u)^2 \mathbf{H}^2$$

$$+ \int \psi^2 h(u) |h'(u)| |f| + \int \psi^2 h(u) |h'(u)| |uF|$$

$$+ \int \psi^2 \left\{ \left(1 + \varepsilon^{-1} \right) h'(u)^2 + \varepsilon^{-1} |h(u) h''(u)| \right\} \left(|\mathbf{g}|^2 + |\mathbf{G}|^2 u^2 \right).$$

1.2. Energy solutions with h a power function. Although there will be technical difficulties to be overcome, we wish to apply inequality (122) with $h = h_\beta$ for $\beta \in \mathbb{R} \setminus \left\{ 0, \frac{1}{2} \right\}$, where $h_\beta(t) = t^\beta$, $t > 0$, to a positive weak solution u of (99) or (101), and in addition, to certain positive weak subsolutions and supersolutions. Noting the important property that

$$\Gamma(t) = \frac{2\beta - 1}{\beta} h_\beta'(t)^2 \begin{cases} > 0 & \text{for} \quad \beta < 0, \beta > \frac{1}{2} \\ < 0 & \text{for} \quad 0 < \beta < \frac{1}{2} \end{cases},$$

we want to apply (122) with $h = h_\beta$ when u is a positive weak $\begin{pmatrix} \text{subsolution} \\ \text{supersolution} \end{pmatrix}$ of (101) and $\begin{pmatrix} \beta > \frac{1}{2} \\ \beta < \frac{1}{2} \end{pmatrix}$. Moreover, in the proof of our local boundedness result, Proposition 59 below, we will need to exploit the more general fact that if $h'(u) \geq 0$, then $\tilde{u}$ is a $\mathcal{E}[\tilde{u}]$-weak subsolution of (117) for a $\mathcal{A}[u]$-weak subsolution u of (101) in Ω, where the subset $\mathcal{A}[u]$ of "admissible test functions" associated to u is given by

$$(123) \quad \mathcal{A}[u] = \left\{ \psi^2 h(u) h'(u) : \psi \in C_0^{0,1}(\Omega), h \text{ is admissible for } u \text{ and } h' \geq 0 \right\}.$$

The notion of $\mathcal{A}[u]$-weak subsolution is more general than weak subsolution since $\mathcal{A}[u] \subset W_0^{1,2}(\Omega)$ by (119). We need the following lemma.

LEMMA 56. *Suppose u is a $\mathcal{A}[u]$-weak subsolution of (101) in Ω, h is admissible for u and $h'(u) \geq 0$. Then $\tilde{u} = h(u)$ is a positive $\mathcal{E}[\tilde{u}]$-weak subsolution of (117) in Ω.*

PROOF (OF LEMMA 56). Just as in Remark 55, we see by examining the first equality in (116) that $\tilde{u} = h(u)$ is a $\mathcal{E}[\tilde{u}]$-weak subsolution of (117) in Ω provided

$$\int \left\{ wh'(u)(Lu + \mathbf{HR}u) \right\} \geq \int \left\{ wh'(u)(f + \mathbf{T}'\mathbf{g} - \mathbf{S}'\mathbf{G}u - Fu) \right\}$$

for all $w \in \mathcal{E}[\tilde{u}]$, i.e. $w = \psi^2 h(u)$ for $\psi \in C_0^{0,1}(\Omega)$. This inequality however, follows from the assumption that u is a $\mathcal{A}[u]$-weak subsolution of (101) in Ω. This completes the proof of Lemma 56.

REMARK 57. *In order to describe appropriate notions of $\mathcal{W}$-weak solution and $\mathcal{W}$-weak sense of equations, we have introduced in (110), (118) and (123) three subsets $\mathcal{W}$ of the nonnegative elements of $W_0^{1,2}(\Omega)$ that are associated to a function $u \in W^{1,2}(\Omega)$ (and in the first case to an admissible h as well):*

$$\mathcal{M}[u; h] = \left\{ w \in W_0^{1,2}(\Omega) : w \geq 0 \text{ and } h'(u)w \in W_0^{1,2}(\Omega) \right\},$$

$$\mathcal{E}[u] = \left\{ \psi^2 u : \psi \in C_0^{0,1}(\Omega) \right\},$$

$$\mathcal{A}[u] = \left\{ \psi^2 h(u) h'(u) : \psi \in C_0^{0,1}(\Omega), h \text{ is admissible for } u \text{ and } h' \geq 0 \right\}.$$

We summarize here the basic properties of these subsets and their relation to standard Moser iteration. If u is a weak solution to (101), then $\mathcal{M}[u; h]$ is the maximal subset $\mathcal{W}$ such that $\tilde{u} = h(u)$ is a $\mathcal{W}$-weak solution to (117). The set of energy test functions $\mathcal{E}[\tilde{u}]$ is the minimal subset $\mathcal{W}$ such that the Caccioppoli inequality (122) holds for $\mathcal{W}$-weak solutions $\tilde{u} = h(u)$ to (117) when $u \in W^{1,2}(\Omega)$ and h is admissible for u. The set of admissible test functions $\mathcal{A}[u]$ is the minimal subset $\mathcal{W}$ such that $\mathcal{W}$-weak solutions u to (101) enjoy the property that $\tilde{u} = h(u)$ is a $\mathcal{E}[\tilde{u}]$-weak solution to (117) for all h that are admissible for u. In standard versions of Moser iteration, one uses test functions $w = \psi^2 h(u) h'(u)$ from the class $\mathcal{A}[u]$ directly in equation (101), with special choices of h. For example, $h(t) = t^\beta$ yields $w = \beta\psi^2 u^{2\beta-1}$ and $h(t) = \sqrt{2 \log \frac{t+m}{m}}$ yields $w = \psi^2(u+m)^{-1}$.

However, as mentioned above, there are difficulties to be overcome in applying inequality (122) with $h = h_\beta$. The difficulties are related to the fact that h_β is not in general admissible for positive u in $W^{1,2}$. We now assume in addition that $u(x)$ is bounded below by a positive constant m. Then for $\beta \leq 1$, h'_β is bounded and Lipschitz on $[m, \infty)$, the range of u. In order to deal with the fact that h'_β fails to be bounded and Lipschitz on $[m, \infty)$ when $\beta > 1$, we use the technique of truncating derivatives of the power function t^β (e.g. as in Theorem 8.15 of [**14**]). For $\beta > 1$ and $M > 0$, let

$$h_{\beta,M}(t) = \begin{cases} t^\beta, & 0 < t \leq M \\ M^\beta + \beta M^{\beta-1}(t-M), & t > M \end{cases},$$

and for convenience in notation, let $h_{\beta,M}(t) = h_\beta(t)$ for $\beta \leq 1$ and $M > 0$. Then the functions $h_{\beta,M}$ are admissible on the intervals $[m, \infty)$ for all $\beta \in \mathbb{R}$, $M > 0$ and $m > 0$.

Motivated by the above concerns, we record that for $u \in W^{1,2}(\Omega)$ and $\widetilde{u} = h_{\beta,M}(u)$,

(a) When $\beta > \frac{1}{2}$ and u is a positive $\mathcal{A}[u]$-weak subsolution of (101), we have that $\Gamma > 0$ and that $\widetilde{u}$ is a $\mathcal{E}[\widetilde{u}]$-weak subsolution of (117) by Lemma 56, since $h'_{\beta,M} > 0$. Thus assumption i above holds.

(b) When $0 < \beta < \frac{1}{2}$ and u is a positive weak supersolution of (101), we have that $\Gamma < 0$ and that $\widetilde{u}$ is a $\mathcal{E}[\widetilde{u}]$-weak supersolution of (117) by Remark 55, since $h'_{\beta,M} > 0$. Thus assumption ii above holds.

(c) When $\beta < 0$ and u is a positive weak supersolution of (101), we have that $\Gamma > 0$ and that $\widetilde{u}$ is a $\mathcal{E}[\widetilde{u}]$-weak subsolution of (117) by Remark 55, since $h'_{\beta,M} < 0$. Thus assumption i above holds.

We summarize our assumptions on u, bounded below by a positive constant, and β, not equal to 0 or $\frac{1}{2}$, as follows - one or more of the following three conditions holds:

$$(124) \quad \begin{cases} u \text{ is a positive weak solution of (101) in } \Omega \\ u \text{ is a positive } \mathcal{A}[u]\text{-weak subsolution of (101) in } \Omega \text{ and } \beta > \frac{1}{2} \\ u \text{ is a positive weak supersolution of (101) in } \Omega \text{ and } \beta < \frac{1}{2} \end{cases}.$$

We could generalize the third assumption in (124) by requiring merely that u be a positive $\mathcal{A}^-[u]$-weak supersolution of (101) in Ω and $\beta < \frac{1}{2}$, where

$$\mathcal{A}^-[u] = \left\{ \psi^2 h(u) h'(u) : \psi \in C_0^{0,1}(\Omega), h \text{ is admissible for } u \text{ and } h' \leq 0 \right\},$$

but will refrain from doing so since we will not need it. We also compute that

$$\Gamma(u) = h'_{\beta,M}(u)^2 + h_{\beta,M}(u) h''_{\beta,M}(u) = \eta_{\beta,M}(u) h'_{\beta,M}(u)^2,$$

where

$$\eta_{\beta,M}(t) = \begin{cases} \frac{2\beta-1}{\beta}, & 0 \leq t \leq M \text{ or } \beta \leq 1 \\ 1, & t > M \text{ and } \beta > 1 \end{cases}.$$

Then by the above discussion, (122) holds with $h = h_{\beta,M}$ when u satisfies (124) since $h'(t)$, $h''(t)$ and $th''(t)$ are all bounded and since the signs of Γ and $h'_{\beta,M}$ are in accordance with (a), (b) and (c) above, so that either (i) or (ii) holds. Now let

$$\mu_\beta = \min\left\{ \left| \frac{2\beta-1}{\beta} \right|, 1 \right\},$$

and choose ε so that $(1 - 2N\varepsilon)\mu_\beta - (5N + 1)\varepsilon = \frac{1}{2}\mu_\beta$. For $\beta \neq \frac{1}{2}$, we have $\mu_\beta > 0$ and $\varepsilon = \frac{\mu_\beta}{10N + 2 + 4N\mu_\beta} > 0$, and we compute that the left-hand side of (122) satisfies

$$\int \psi^2 \left[(1 - 2N\varepsilon)\left|\Gamma(u)\right| - (5N + 1)\varepsilon h'(u)^2 \right] \|\nabla u\|^2$$

$$= \int \psi^2 \left[(1 - 2N\varepsilon)\left|\eta_{\beta,M}(u)\right| - (5N + 1)\varepsilon \right] h'_{\beta,M}(u)^2 \|\nabla u\|^2$$

$$\geq \int \psi^2 \left[(1 - 2N\varepsilon)\mu_\beta - (5N + 1)\varepsilon \right] h'_{\beta,M}(u)^2 \|\nabla u\|^2$$

$$= \frac{1}{2}\mu_\beta \int \psi^2 h'_{\beta,M}(u)^2 \|\nabla u\|^2.$$

Thus (122) yields

$$(125) \quad \frac{1}{2}\mu_\beta \int \psi^2 h'_{\beta,M}(u)^2 \|\nabla u\|^2$$

$$\leq C\left(1 + \mu_\beta^{-1}\right) \int h_{\beta,M}(u)^2 \left\{ \|\nabla\psi\|^2 + \psi^2 |\mathbf{H}|^2 \right\}$$

$$+ \int \psi^2 h_{\beta,M}(u)\left|h'_{\beta,M}(u)\right|(|f| + u|F|)$$

$$+ C\left(1 + \mu_\beta^{-1}\right) \int \psi^2 \left\{ h'_{\beta,M}(u)^2 + \left|h_{\beta,M}(u)h''_{\beta,M}(u)\right| \right\}\left(|\mathbf{g}|^2 + u^2|\mathbf{G}|^2\right).$$

Now let $M \to \infty$ in (125) to obtain that (125) holds with h_β in place of $h_{\beta,M}$. To see this, we note that for $\beta > 1$, each of the functions $h_{\beta,M}^2$, $\left|h_{\beta,M}h'_{\beta,M}\right|$, $\left(h'_{\beta,M}\right)^2$ and $\left|h_{\beta,M}(u)h''_{\beta,M}(u)\right|$ increases as M increases to ∞, and so the monotone convergence theorem applies.

Altogether then, since $\int \psi^2 h'_\beta(u)^2 \|\nabla u\|^2 = \int \psi^2 \|\nabla u^\beta\|^2$, we have

$$(126) \quad \int \psi^2 \|\nabla u^\beta\|^2$$

$$\leq C\left(1 + \mu_\beta^{-2}\right) \int u^{2\beta}\left\{ \|\nabla\psi\|^2 + \psi^2 |\mathbf{H}|^2 \right\}$$

$$+ 2\mu_\beta^{-1}|\beta| \int \psi^2 u^{2\beta - 1}(|f| + u|F|)$$

$$+ C\left(\beta^2 + |\beta|\right)\left(1 + \mu_\beta^{-2}\right) \int \psi^2 u^{2\beta - 2}\left(|\mathbf{g}|^2 + u^2|\mathbf{G}|^2\right),$$

whenever $\psi \in C_0^{0,1}(\Omega)$, $\beta \in \mathbb{R}$ with $\beta \neq 0, \frac{1}{2}$, and u is bounded below by a positive constant and satisfies (124). Note that we are not assuming that the right side of (126) is finite at this point.

1.3. Inhomogeneous equations. In order to handle the terms involving f and $\mathbf{g}$ on the right side of (126), as well as allowing u to be merely nonnegative, we will want to apply inequality (126) (with appropriate restrictions on β) to a *positive*

$$\begin{pmatrix} \text{weak solution} \\ A\left[\overline{u}\right]\text{-weak subsolution} \\ \text{weak supersolution} \end{pmatrix} \overline{u} = u + m(r) \text{ of the equation}$$

$$(127) \qquad \qquad \mathcal{L}\overline{u} = f + m(r)F + \mathbf{T}'\mathbf{g} + \mathbf{S}'m(r)\mathbf{G}$$

in Ω, where $\mathcal{L} = L + \mathbf{H}\mathbf{R} + \mathbf{S}'\mathbf{G} + F$ and

$$(128) \qquad m(r) = m_{q,\eta}(r) = r^{2\eta}\|f\|_{\frac{q}{2}} + r^{\eta}\|\mathbf{g}\|_q,$$

and $\eta > 0$ satisfies $q(1-\eta) > Q$. The result of this application will be the following variant of (126):

$$(129) \qquad \int \psi^2 \left\|\nabla \overline{u}^{\beta}\right\|^2$$

$$\leq \; C\left(1 + \mu_{\beta}^{-2}\right) \int \overline{u}^{2\beta} \left\{\|\nabla\psi\|^2 + \psi^2 |\mathbf{H}|^2\right\}$$

$$+ 2\mu_{\beta}^{-1}|\beta| \int \psi^2 \overline{u}^{2\beta-1}\left(|f + m(r)F| + \overline{u}|F|\right)$$

$$+ C\left(\beta^2 + |\beta|\right)\left(1 + \mu_{\beta}^{-2}\right) \int \psi^2 \overline{u}^{2\beta-2}\left(|\mathbf{g}|^2 + |m(r)\mathbf{G}|^2 + \overline{u}^2 |\mathbf{G}|^2\right).$$

In the case $m(r) = 0$, we use $m > 0$ and then let $m \to 0$ at the end of the argument, thus ensuring that at all times, $\overline{u}$ is bounded below by a positive constant. Now if u satisfies

$$\begin{cases} u \text{ is a nonnegative weak solution of (101) in } \Omega \\ u \text{ is a nonnegative weak subsolution of (101) in } \Omega \text{ and } \beta > \frac{1}{2} \\ u \text{ is a nonnegative weak supersolution of (101) in } \Omega \text{ and } \beta < \frac{1}{2} \end{cases},$$

then it is easy to see that

$$\begin{cases} \overline{u} \text{ is a positive weak solution of (127) in } \Omega \\ \overline{u} \text{ is a positive weak subsolution of (127) in } \Omega \text{ and } \beta > \frac{1}{2} \\ \overline{u} \text{ is a positive weak supersolution of (127) in } \Omega \text{ and } \beta < \frac{1}{2} \end{cases}.$$

However, if u merely satisfies (124), then a problem arises with the middle assumption there, namely that if u is a positive $\mathcal{A}[u]$-weak subsolution of (101) in Ω, it does not necessarily follow that $\overline{u}$ is a positive $\mathcal{A}[\overline{u}]$-weak subsolution of (127) in Ω. The difficulty is that the collection of test functions changes from $\mathcal{A}[u]$ to $\mathcal{A}[\overline{u}]$. Thus for $m > 0$, we consider the equation

$$(130) \qquad \mathcal{L}(u + m) = f + mF + \mathbf{T}'\mathbf{g} + \mathbf{S}'m\mathbf{G},$$

and will now assume that u satisfies one of the following four assumptions with $\beta \neq 0, \frac{1}{2}$:

$$(131) \qquad \begin{cases} u \text{ is a nonnegative weak solution of (101) in } \Omega \\ u \text{ is a nonnegative weak subsolution of (101) in } \Omega \text{ and } \beta > \frac{1}{2} \\ u + m \text{ is a positive } \mathcal{A}[u+m]\text{-weak subsolution of (130) in } \Omega \\ \quad \text{for all } m > 0, \text{ and } \beta > \frac{1}{2} \\ u \text{ is a nonnegative weak supersolution of (101) in } \Omega \text{ and } \beta < \frac{1}{2} \end{cases}.$$

We note that the second assumption in (131) implies the third since if u is a nonnegative weak subsolution of (101), then $u + m$ is a positive weak subsolution of (130), and then $u + m$ is a positive $\mathcal{A}[u+m]$-weak subsolution of (130) since $\mathcal{A}[u+m] \subset W_0^{1,2}(\Omega)$ by (119). We are including the second assumption in (131) explicitly so as to emphasize the third, which is used only in the proof of Proposition 59. Note that if the third condition in (131) holds, then the choice $m = m(r)$ shows that $\overline{u}$ is a $\mathcal{A}[\overline{u}]$-weak subsolution of (127), and hence that (129) holds as desired.

Using $\bar{u}^{-1} \leq m(r)^{-1}$, we obtain from (129) that for $\beta \neq 0, \frac{1}{2}$,

$$\int \psi^2 \left\| \nabla \bar{u}^\beta \right\|^2$$

$$\leq C \left(1 + \mu_\beta^{-2} \right) \int \bar{u}^{2\beta} \left\{ \left\| \nabla \psi \right\|^2 + \psi^2 \left| \mathbf{H} \right|^2 \right\}$$

$$+ 2 \mu_\beta^{-1} |\beta| \int \psi^2 \bar{u}^{2\beta} \left\{ m(r)^{-1} |f + m(r) F| + |F| \right\}$$

$$+ C \left(\beta^2 + |\beta| \right) \left(1 + \mu_\beta^{-2} \right) \int \psi^2 \bar{u}^{2\beta} \left\{ \left| m(r)^{-1} \mathbf{g} \right|^2 + |\mathbf{G}|^2 \right\},$$

where for the last line we have used

$$u^{-2} \left(|\mathbf{g}|^2 + m(r)^2 |\mathbf{G}|^2 + u^2 |\mathbf{G}|^2 \right) \leq \left| m(r)^{-1} \mathbf{g} \right|^2 + 2 |\mathbf{G}|^2.$$

Using Hölder's inequality with exponents $\frac{q}{2}$ and $\left(\frac{q}{2}\right)'$, we thus have

$$(132) \qquad \int \psi^2 \left\| \nabla \bar{u}^\beta \right\|^2$$

$$\leq C_\beta \int \bar{u}^{2\beta} \left\| \nabla \psi \right\|^2 + C_\beta \left\| \mathbf{H} \right\|_q^2 \left(\int \left(\psi^2 \bar{u}^{2\beta} \right)^{\left(\frac{q}{2}\right)'} \right)^{\frac{1}{\left(\frac{q}{2}\right)'}}$$

$$+ C_\beta \left\{ m(r)^{-1} \left\| f \right\|_{\frac{q}{2}} + \left(m(r)^{-1} \left\| \mathbf{g} \right\|_q \right)^2 \right\} \left(\int \left(\psi^2 \bar{u}^{2\beta} \right)^{\left(\frac{q}{2}\right)'} \right)^{\frac{1}{\left(\frac{q}{2}\right)'}}$$

$$+ C_\beta \left\{ \left\| F \right\|_{\frac{q}{2}} + \left\| \mathbf{G} \right\|_q^2 \right\} \left(\int \left(\psi^2 \bar{u}^{2\beta} \right)^{\left(\frac{q}{2}\right)'} \right)^{\frac{1}{\left(\frac{q}{2}\right)'}}$$

$$\leq C_\beta \int \bar{u}^{2\beta} \left\| \nabla \psi \right\|^2 + C_\beta \left(1 + r^{-2\eta} \right) \left(\int \left(\psi^2 \bar{u}^{2\beta} \right)^{\left(\frac{q}{2}\right)'} \right)^{\frac{1}{\left(\frac{q}{2}\right)'}},$$

where the constant C_β is dominated by $C \Upsilon(\beta)$ for the following positive function Υ of β that blows up at $0, \frac{1}{2}$ and ∞,

$$(133) \qquad \Upsilon(\beta) = \left(|\beta| + \frac{1}{|\beta|} + \frac{1}{|\beta - \frac{1}{2}|} \right)^\tau,$$

and where C depends on q, $\|F\|_{\frac{q}{2}}$, $\|\mathbf{G}\|_q$ and $\|\mathbf{H}\|_q$, but is *independent* of u, r, β, $\|f\|_{\frac{q}{2}}$ and $\|\mathbf{g}\|_q$, and where τ is positive and *independent* of u, r, β, F, $\mathbf{G}$, $\mathbf{H}$, f, $\mathbf{g}$ and q, but may vary from line to line. Note again that we are not assuming the right side of (132) to be finite.

2. Weak reverse Hölder inequalities and Moser iteration

We now invoke the homogeneous space structure given by the quasimetric d in the hypotheses of Theorem 8. Let u be a *nonnegative* weak solution of (101), or more generally satisfy (131), so that $\bar{u}$ *is* positive (see the discussion surrounding (128)). To implement Moser iteration, we fix a d-ball $B = B(y, r)$ with $y \in \Omega$, $0 < r < \delta \, dist(y, \partial \Omega)$, and consider the sequence of Lipschitz functions ψ_j given in (20), along with the sets $E_j = supp \, \psi_j$. Note that since $B(y, cr) \subset E_j \subset B(y, r)$, we have $|E_j| \approx |B|$ by doubling. Fix $\beta \neq 0, \frac{1}{2}$. If we substitute ψ_j for ψ in (132), divide through by $|E_j|$, and then take square roots, we obtain

(134)
$$\left\{ \frac{1}{|E_j|} \int \psi_j^2 \left\| \nabla \overline{u}^\beta \right\|^2 \right\}^{\frac{1}{2}}$$

$$\leq C\Upsilon(\beta) \left\{ \frac{1}{|E_j|} \int \overline{u}^{2\beta} \left\| \nabla \psi_j \right\|^2 \right\}^{\frac{1}{2}}$$

$$+ C\Upsilon(\beta) \left(1 + r^{-\eta} \right) |E_j|^{\frac{1}{2\left(\frac{q}{2}\right)'} - \frac{1}{2}} \left\{ \frac{1}{|E_j|} \int \left(\psi_j^2 \overline{u}^{2\beta} \right)^{\left(\frac{q}{2}\right)'} \right\}^{\frac{1}{2\left(\frac{q}{2}\right)'}}$$

$$\leq C\Upsilon(\beta) \frac{j^N}{r} \left\{ \frac{1}{|E_j|} \int_{E_j} \overline{u}^{2\beta\mu} \right\}^{\frac{1}{2\mu}} + C\Upsilon(\beta) \frac{1}{r} \left\{ \frac{1}{|E_j|} \int \left(\psi_j^2 \overline{u}^{2\beta} \right)^{\nu} \right\}^{\frac{1}{2\nu}},$$

with

(135)
$$\mu = \left(\frac{p}{2} \right)', \nu = \left(\frac{q}{2} \right)'.$$

Indeed, for $q > q(1 - \eta) > Q$, (10) implies both $|E_j| \geq |B(y, cr)| \geq c_{q(1-\eta)} (cr)^{q(1-\eta)}$ and $|E_j| \geq c_q (cr)^q$, and so

$$\left(1 + r^{-\eta} \right) |E_j|^{\frac{1}{2\left(\frac{q}{2}\right)'} - \frac{1}{2}} = \left(1 + r^{-\eta} \right) |E_j|^{-\frac{1}{q}} \leq c^{-1} c_q^{-\frac{1}{q}} r^{-1} + c^{\eta-1} c_{q(1-\eta)}^{-\frac{1}{q}} r^{-1}.$$

Moreover,

(136)
$$\frac{1}{|E_j|} \int \overline{u}^{2\beta} \left\| \nabla \psi_j \right\|^2 \leq \left\{ \frac{1}{|E_j|} \int \left\| \nabla \psi_j \right\|^p dx \right\}^{\frac{2}{p}} \left\{ \frac{1}{|E_j|} \int_{E_j} \overline{u}^{2\beta\left(\frac{p}{2}\right)'} \right\}^{\frac{1}{\left(\frac{p}{2}\right)'}}$$

$$\leq C_{sym} C_p^2 \left| \frac{j^N}{r} \right|^2 \left\{ \frac{1}{|E_j|} \int_{E_j} \overline{u}^{2\beta\mu} \right\}^{\frac{1}{\mu}},$$

upon using Hölder's inequality with exponents $\frac{p}{2}$ and $\left(\frac{p}{2} \right)'$, followed by an application of (20) and (27).

We now use the Sobolev inequality (15) to obtain that for some $\sigma > 1$ and any $w \in W^{1,2}(B)$,

(137)
$$\left\{ \frac{1}{|B|} \int_B |\psi_j w|^{2\sigma} \right\}^{\frac{1}{2\sigma}}$$

$$\leq \frac{1}{\sqrt{c_{sym}}} Cr \left\{ \frac{1}{|B|} \int_B \left\| \nabla(\psi_j w) \right\|^2 \right\}^{\frac{1}{2}} + C \left\{ \frac{1}{|B|} \int_B |\psi_j w|^2 \right\}^{\frac{1}{2}}$$

$$\leq Cr \left\{ \frac{1}{|B|} \int_B \psi_j^2 \left\| \nabla w \right\|^2 \right\}^{\frac{1}{2}} + C \left\{ \frac{1}{|B|} \int_B \left(\psi_j^2 + r^2 \left\| \nabla \psi_j \right\|^2 \right) |w|^2 \right\}^{\frac{1}{2}},$$

for $j \geq 1$. For $\beta \leq 1$, $\overline{u}^\beta \in W^{1,2}(B)$ (recall that $\overline{u}$ is bounded below by a positive constant) and thus (137) holds with $w = \overline{u}^\beta$. For $\beta > 1$, we have (137) for $w = h_{\beta,M}(\overline{u})$, and the monotone convergence theorem then yields (137) for $w = \overline{u}^\beta$ upon letting $M \to \infty$. Using $|E_j| \approx |B|$ and combining (134) and (137) with $w = \overline{u}^\beta$ we obtain

$$
\left\{\frac{1}{|B|}\int_B \left(\psi_j \bar{u}^\beta\right)^{2\sigma}\right\}^{\frac{1}{2\sigma}}
$$

$$
\leq\ Cr\left\{\frac{1}{|B|}\int_B \psi_j^2 \left\|\nabla \bar{u}^\beta\right\|^2\right\}^{\frac{1}{2}}
$$

$$
+C\left\{\frac{1}{|B|}\int_B \left(\psi_j^2 + r^2 \left\|\nabla \psi_j\right\|^2\right)\left|\bar{u}^\beta\right|^2\right\}^{\frac{1}{2}}
$$

$$
\leq\ C\Upsilon(\beta)\, j^N \left\{\frac{1}{|E_j|}\int_{E_j} \bar{u}^{2\beta\mu}\right\}^{\frac{1}{2\mu}} + C\left(\Upsilon(\beta)+1\right)\left\{\frac{1}{|E_j|}\int \left(\psi_j \bar{u}^\beta\right)^{2\nu}\right\}^{\frac{1}{2\nu}}
$$

$$
+Cr\left\{\frac{1}{|E_j|}\int_{E_j} \left\|\nabla \psi_j\right\|^2 \bar{u}^{2\beta}\right\}^{\frac{1}{2}},
$$

and hence using (136) again on the final term,

$$
(138)\quad \left\{\frac{1}{|E_{j+1}|}\int \left(\psi_j \bar{u}^\beta\right)^{2\sigma}\right\}^{\frac{1}{2\sigma}} \leq\ C\Upsilon(\beta)\, j^N \left\{\frac{1}{|E_j|}\int_{E_j} \bar{u}^{2\beta\mu}\right\}^{\frac{1}{2\mu}}
$$

$$
+C\left(\Upsilon(\beta)+1\right)\left\{\frac{1}{|E_j|}\int \left(\psi_j \bar{u}^\beta\right)^{2\nu}\right\}^{\frac{1}{2\nu}}
$$

for $j \geq 1$, where the constant C depends on c_{sym}, C_{sym}, p, q, $\|F\|_{\frac{q}{2}}$, $\|\mathbf{G}\|_q$ and $\|\mathbf{H}\|_q$, but is *independent* of u, r, β, $\|f\|_{\frac{q}{2}}$ and $\|\mathbf{g}\|_q$. Recall now that $\max\{\mu,\nu\} < \sigma$ by hypothesis, and thus in particular, with

$$
(139)\qquad\qquad\qquad \max\{\mu,\nu\} \leq \rho < \sigma,
$$

we have

$$
(140)\qquad \left\{\frac{1}{|E_{j+1}|}\int \left(\psi_j \bar{u}^\beta\right)^{2\sigma}\right\}^{\frac{1}{2\sigma}} \leq C\Upsilon(\beta)\, j^N \left\{\frac{1}{|E_j|}\int_{E_j} \bar{u}^{2\beta\rho}\right\}^{\frac{1}{2\rho}}.
$$

In the case $\nu \leq \mu$, the second term on the right of (138) is majorized by the first term on the right by Hölder's inequality. We now interrupt the main line of our development of Moser iteration to point out how to majorize the second term by the first when $\mu < \nu$ and Hölder's inequality is unavailable. This will only be needed in the case $\beta > \frac{1}{2}$ and we will appeal to the Lebesgue space interpolation inequality (see e.g. (7.10) of Chapter 7 of [**14**])

$$
(141)\quad \|w\|_{L^s(d\lambda)} \leq \varepsilon \|w\|_{L^t(d\lambda)} + \left(\frac{1}{\varepsilon}\right)^{\frac{\left(\frac{1}{r}-\frac{1}{s}\right)}{\left(\frac{1}{s}-\frac{1}{t}\right)}} \|w\|_{L^r(d\lambda)},\quad r < s < t, \varepsilon > 0,
$$

valid for any positive measure $d\lambda$, in order to estimate the final integral in (138). We have $\nu < \sigma$ by hypothesis and we now assume $\mu < \nu$ and $\beta > \frac{1}{2}$. Let $r = 2\mu$, $s = 2\nu$, $t = 2\sigma$ and $w = \psi_j \bar{u}^\beta$ in (141). Then (141) with $d\lambda = \frac{1}{|E_j|}\chi_{E_j}(x)\, dx$ and

(138) yield

$$\left\{ \frac{1}{|E_j|} \int \left(\psi_j \overline{u}^\beta \right)^{2\nu} \right\}^{\frac{1}{2\nu}}$$

$$\leq \varepsilon \left\{ \frac{1}{|E_j|} \int \left(\psi_j \overline{u}^\beta \right)^{2\sigma} \right\}^{\frac{1}{2\sigma}} + \left(\frac{1}{\varepsilon} \right)^{\frac{\left(\frac{1}{\mu} - \frac{1}{\nu} \right)}{\left(\frac{1}{\nu} - \frac{1}{\sigma} \right)}} \left\{ \frac{1}{|E_j|} \int \left(\psi_j \overline{u}^\beta \right)^{2\mu} \right\}^{\frac{1}{2\mu}}$$

$$\leq \varepsilon C \Upsilon \left(\beta \right) j^N \left\{ \frac{1}{|E_j|} \int_{E_j} \overline{u}^{2\beta\mu} \right\}^{\frac{1}{2\mu}} + \varepsilon C \left(\Upsilon \left(\beta \right) + 1 \right) \left\{ \frac{1}{|E_j|} \int \left(\psi_j \overline{u}^\beta \right)^{2\nu} \right\}^{\frac{1}{2\nu}}$$

$$+ \left(\frac{1}{\varepsilon} \right)^{\frac{\left(\frac{1}{\mu} - \frac{1}{\nu} \right)}{\left(\frac{1}{\nu} - \frac{1}{\sigma} \right)}} \left\{ \frac{1}{|E_j|} \int_{E_j} \overline{u}^{2\beta\mu} \right\}^{\frac{1}{2\mu}} .$$

Now choose ε so that $\varepsilon C \left(\Upsilon \left(\beta \right) + 1 \right) = \frac{1}{2}$. Provided the integral $\int \left(\psi_j \overline{u}^\beta \right)^{2\nu}$ is finite, we can absorb the second term on the right into the left-hand side and obtain

$$(142) \qquad \left\{ \frac{1}{|E_j|} \int \left(\psi_j \overline{u}^\beta \right)^{2\nu} \right\}^{\frac{1}{2\nu}} \leq C \left(p, q, \sigma, \beta, j^N \right) \left\{ \frac{1}{|E_j|} \int_{E_j} \overline{u}^{2\beta\mu} \right\}^{\frac{1}{2\mu}} ,$$

(where $C \left(p, q, \sigma, \beta, j^N \right)$ depends also on $\|F\|_{\frac{q}{2}}$, $\|\mathbf{G}\|_q$ and $\|\mathbf{H}\|_q$). To prove (142), we have assumed $\beta > \frac{1}{2}$, $\int_{E_j} \overline{u}^{2\beta\nu}$ is finite, u satisfies (131) and that $B \left(y, cr \right) \subset E_j \subset B \left(y, r \right)$ as above.

Now we return to the main line of our development of Moser iteration. For the moment, fix ρ satisfying (139) and let $\theta = \frac{\sigma}{\rho}$ so that $\theta > 1$. Fix $\gamma \neq 0$ and real and let $u_j = \overline{u}^{\gamma \theta^{j-1}}$. Using $u_{j+1} = u_j^\theta$, $\psi_j = 1$ on E_{j+1} and applying (140) with $\overline{u}^{\beta\rho} = u_j$, i.e. $\beta = \frac{\gamma \theta^{j-1}}{\rho} = \frac{\gamma \sigma^{j-1}}{\rho^j}$, yields

$$(143) \qquad \left\{ \frac{1}{|E_{j+1}|} \int_{E_{j+1}} |u_{j+1}|^2 \right\}^{\frac{1}{2}} \leq C \Upsilon \left(\frac{\gamma \theta^{j-1}}{\rho} \right) j^{N\sigma} \left\{ \frac{1}{|E_j|} \int_{E_j} |u_j|^2 \right\}^{\frac{\theta}{2}} .$$

Since (140) requires (131), we assume for the above that one of the following holds:
(144)
$$\begin{cases} u \text{ is a nonnegative weak solution of (101) in } \Omega \text{ and } \gamma \neq 0, \frac{\rho}{2\theta^{j-1}} \\ u + m \text{ is a positive } \mathcal{A} \left[u + m \right] \text{-weak subsolution of (130) in } \Omega \\ \quad \text{for all } m > 0, \text{ and } \gamma > \frac{\rho}{2\theta^{j-1}} \\ u \text{ is a nonnegative weak supersolution of (101) in } \Omega \text{ and } \gamma < \frac{\rho}{2\theta^{j-1}}, \gamma \neq 0 \end{cases} .$$

Recall from the discussion following (131) that the second assumption there implies the third. Set

$$\mathbf{N}_j = \left\{ \frac{1}{|E_j|} \int_{E_j} |\overline{u}|^{2\gamma \theta^{j-1}} \right\}^{\frac{1}{2\theta^{j-1}}} = \left\{ \frac{1}{|E_j|} \int_{E_j} |u_j|^2 \right\}^{\frac{1}{2\theta^{j-1}}} ,$$

for $j \geq 1$, so that if we raise (143) to the power $\frac{1}{\theta^j}$, we obtain

$$(145) \qquad \mathbf{N}_{j+1} \leq \left[C \Upsilon \left(\frac{\gamma \theta^{j-1}}{\rho} \right) j^{N\sigma} \right]^{\frac{1}{\theta^j}} \mathbf{N}_j \quad \text{provided (144) holds.}$$

In order to iterate (145) for $j \geq 1$, we must have that (144) holds for *all* $j \geq 1$ for some choice of ρ satisfying (139). It turns out that with an appropriate choice of ρ it is enough to assume that one of the following holds:

$$(146) \quad \begin{cases} u \text{ is a nonnegative weak solution of (101) in } \Omega \text{ and } \gamma \neq 0 \\ u + m \text{ is a positive } \mathcal{A}\left[u + m\right]\text{-weak subsolution of (130) in } \Omega \\ \text{for all } m > 0, \text{and } \gamma > \frac{1}{2}\max\{\mu, \nu\} \\ u \text{ is a nonnegative weak supersolution of (101) in } \Omega \text{ and } \gamma < 0 \end{cases}.$$

Indeed, we claim that if (146) holds, then with c as in (20),

$$(147) \quad \operatorname*{ess\,sup}_{x \in B(y, cr)} |\overline{u}(x)|^{\gamma} \leq \limsup_{j \to \infty} \mathbf{N}_j \leq C_{\sigma, \gamma, \tau} \left\{ \frac{1}{|B(y, r)|} \int_{B(y, r)} |\overline{u}|^{2\gamma} \right\}^{\frac{1}{2}},$$

where the constant $C_{\sigma, \gamma, \tau}$ depends as well on p, q, $\|F\|_{\frac{q}{2}}$, $\|\mathbf{G}\|_q$ and $\|\mathbf{H}\|_q$, but is *independent* of u, $B(y, r)$, $\|f\|_{\frac{q}{2}}$ and $\|\mathbf{g}\|_q$ (τ is as in (133)). The first inequality in (147) is standard and for the second, (145) yields

$$\limsup_{j \to \infty} \mathbf{N}_j \leq \left\{ \prod_{j=1}^{\infty} \left[C\Upsilon\left(\frac{\gamma \theta^{j-1}}{\rho} \right) j^{N\sigma} \right]^{\frac{1}{\theta^j}} \right\} \mathbf{N}_1$$

$$\leq C_{\sigma, \rho, \gamma, \tau} \left\{ \frac{1}{|B(y, r)|} \int_{B(y, r)} |\overline{u}|^{2\gamma} \right\}^{\frac{1}{2}},$$

with

$$C_{\sigma, \rho, \gamma, \tau} = \exp \sum_{j=1}^{\infty} \frac{\ln\left[C\left(\left| \frac{\gamma \sigma^{j-1}}{\rho^j} \right| + \left| \frac{\gamma \sigma^{j-1}}{\rho^j} \right|^{-1} + \left| \frac{\gamma \sigma^{j-1}}{\rho^j} - \frac{1}{2} \right|^{-1} \right)^{\tau} j^{N\sigma} \right]}{\sigma^j / \rho^j} < \infty.$$

To ensure that $C_{\sigma, \rho, \gamma, \tau}$ is finite, we must avoid having $\frac{\gamma \sigma^{j-1}}{\rho^j} = \frac{1}{2}$ for any $j = 1, 2, 3, \ldots$, as well as $\gamma \neq 0$. In fact, given $\gamma \neq 0$ and $\sigma > 1$, we now choose ρ in (139) so that the distance of the number $\frac{1}{2}$ from the geometric sequence $\left\{ \frac{\gamma \sigma^{j-1}}{\rho^j} \right\}_{j=1}^{\infty}$ (which depends continuously on ρ) is close to a maximum, thus obtaining

$$\limsup_{j \to \infty} \mathbf{N}_j \leq C_{\sigma, \gamma, \tau} \left\{ \frac{1}{|B(y, r)|} \int_{B(y, r)} |\overline{u}|^{2\gamma} \right\}^{\frac{1}{2}}.$$

This completes the proof that (147) holds whenever (146) holds.

REMARK 58. *The restriction $\gamma > \frac{1}{2}\max\{\mu, \nu\}$ in (146) when $u + m$ is a positive $\mathcal{A}\left[u + m\right]$-weak subsolution for all $m > 0$ will prove problematic in Proposition 59, our local boundedness result, since we will need to choose $\gamma = 1$ in order to obtain finiteness of the right-hand side. When $q \leq 4$ in (24) and (23) (which can only occur in dimensions 2 and 3 since $q \geq Q \geq n$), then $\max\{\mu, \nu\} \geq \nu = \left(\frac{q}{2} \right)' \geq 2$ and it is not true that $\gamma > \frac{1}{2}\max\{\mu, \nu\}$ in case $\gamma = 1$. It is here that we will use (142) (a consequence of the interpolation inequality) for $j = 1$ in concert with the Sobolev inequality (15) to obtain the case $\gamma = 1$ of (147). See the end of the proof of Proposition 59 for this argument. The corresponding problem arises when $p \leq 4$ in (20), but cannot be handled by the interpolation inequality. This accounts for the hypothesis $p > 4$ in Proposition 59 below.*

We can now obtain the local boundedness properties of weak sub and super-solutions to (101) under natural hypotheses. Note that there exist weak subsolutions of Laplace's equation in the unit disk that fail to be bounded below, such as $u(x, y) = -\log(-\log\sqrt{x^2 + y^2})$. For any function u, we use the notations $u_+ = \max\{u, 0\}$ and $u_- = \max\{-u, 0\}$.

PROPOSITION 59. *Suppose that u is a weak solution to (99) in Ω, and that both (23) and (24) hold, as well as (15), (17) and (20) with $p > 4$. Then u is locally bounded in Ω, i.e., for every compact set $K \subset \Omega$,*

$$\|u\|_{L^\infty(K)} \leq C'_K \left(1 + \|u\|_{L^2(\Omega)}\right).$$

More precisely, there is a positive constant c_1 depending on the constant in (20), such that if $y \in \Omega$, $0 < r < \delta \, \text{dist}\,(y, \partial\Omega)$ for sufficiently small δ, then

$$(148) \quad \|u\|_{L^\infty(B(y, c_1 r))} \leq C \left\{ \left(\frac{1}{|B(y, r)|} \int_{B(y,r)} u^2 \right)^{\frac{1}{2}} + r^{2\eta} \|f\|_{\frac{q}{2}} + r^\eta \|\mathbf{g}\|_q \right\},$$

with a constant C that is independent of $\|f\|_{\frac{q}{2}}$ and $\|\mathbf{g}\|_q$. Of course C depends upon the constants in (20) and (15). More generally, if u is a weak subsolution (supersolution) to (99) in Ω, then under the above conditions u is locally bounded above (below) and (148) holds with u replaced on both sides by $u_+(u_-)$.

PROOF. Suppose first that u is a weak subsolution to (99), which is the same as (101), in Ω. Let $m(r) = r^{2\eta} \|f\|_{\frac{q}{2}} + r^\eta \|\mathbf{g}\|_q$ as in (128) (if $m(r) = 0$, replace $m(r)$ with $m > 0$ and let $m \to 0$ at the end of the argument), and set $h(t) = \sqrt{m(r)^2 + t^2}$ for $t \geq 0$ and $h(t) = m(r)$ for $t \leq 0$. Then h is admissible on $\mathbb{R}$ (we remind the reader that the constant C in (108) is only qualitative, and doesn't appear in any estimates for u) and $h' \geq 0$, so that by Remark 55, $\widetilde{u} = h(u)$ is a positive $\mathcal{M}[u; h]$-weak subsolution of (117) in Ω, where $\mathcal{M}[u; h]$ is as in (110), i.e.

$$\mathcal{M}[u; h] = \left\{ w \in W_0^{1,2}(\Omega) : w \geq 0 \text{ and } h'(u) w \in W_0^{1,2}(\Omega) \right\}.$$

Since the vector fields $\mathbf{T}$ and $\mathbf{S}$ are subunit and $h'' \geq 0$, we have the inequality

$$
\begin{aligned}
&h''(u) \|\nabla u\|^2 + \Phi \\
={}& h''(u) \left\{ \|\nabla u\|^2 + (\mathbf{T} u)\, \mathbf{g} - (\mathbf{S} u)\, \mathbf{G} u \right\} \\
\geq{}& h''(u) \left\{ \|\nabla u\|^2 - \varepsilon (\mathbf{T} u)^2 - \frac{1}{\varepsilon} |\mathbf{g}|^2 - \varepsilon (\mathbf{S} u)^2 - \frac{1}{\varepsilon} |\mathbf{G}|^2 u^2 \right\} \\
\geq{}& -C_0 h''(u) \left\{ |\mathbf{g}|^2 + |\mathbf{G}|^2 u^2 \right\}
\end{aligned}
$$

in the $\mathcal{M}[u; h]$-weak sense. Thus $\widetilde{u}(x) = h(u(x))$ is a (positive) $\mathcal{M}[u; h]$-weak subsolution of the equation

$$(149) \quad L\widetilde{u} + \mathbf{H} R\widetilde{u} + \mathbf{S}' \left(\widetilde{\mathbf{G}} \widetilde{u} \right) + \widetilde{F} \widetilde{u} = \widetilde{f} + \mathbf{T}' \widetilde{\mathbf{g}},$$

where

$$\widetilde{\widetilde{F}} = \frac{u}{h(u)}h'(u)F + C_0\frac{u^2}{h(u)}h''(u)|\mathbf{G}|^2,$$

$$\widetilde{\mathbf{G}} = \frac{u}{h(u)}h'(u)\mathbf{G},$$

$$\widetilde{\widetilde{f}} = h'(u)f - C_0h''(u)|\mathbf{g}|^2,$$

$$\widetilde{\mathbf{g}} = h'(u)\mathbf{g}.$$

Now we have

$$\left|\frac{u}{h(u)}h'(u)\right| \leq 1, \left|\frac{u^2}{h(u)}h''(u)\right| \leq 1, |h'(u)| \leq 1,$$

$$|h''(u)| \leq m(r)^{-1},$$

and it follows that

$$\left\|\widetilde{\widetilde{F}}\right\|_{\frac{q}{2}} \leq \|F\|_{\frac{q}{2}} + C_0\|\mathbf{G}\|_q^2,$$

$$\left\|\widetilde{\mathbf{G}}\right\|_q \leq \|\mathbf{G}\|_q,$$

$$\left\|\widetilde{\widetilde{f}}\right\|_{\frac{q}{2}} \leq \|f\|_{\frac{q}{2}} + C_0 m(r)^{-1}\|\mathbf{g}\|_q^2,$$

$$\|\widetilde{\mathbf{g}}\|_q \leq \|\mathbf{g}\|_q.$$

We would like to apply inequality (147) to the (positive) $\mathcal{M}[u;h]$-weak subsolution $\widetilde{u} = h(u)$ of (149), but this requires the middle case of (146), i.e. that $\widetilde{u} + m$ is a positive $\mathcal{A}[\widetilde{u} + m]$-weak subsolution of
(150)

$$L(\widetilde{u}+m)+\mathbf{HR}(\widetilde{u}+m)+\mathbf{S}'\left(\widetilde{\mathbf{G}}(\widetilde{u}+m)\right)+\widetilde{\widetilde{F}}(\widetilde{u}+m) = \left(\widetilde{\widetilde{f}}+m\widetilde{\widetilde{F}}\right)+\mathbf{T}'\widetilde{\mathbf{g}}+\mathbf{S}'m\widetilde{\mathbf{G}}$$

in Ω for all $m > 0$, and $\gamma > \frac{1}{2}\max\{\mu,\nu\}$. Since $\widetilde{u} + m$ is already a subsolution of (150) in the $\mathcal{M}[u;h]$-weak sense, it is enough to establish that $\mathcal{A}[\widetilde{u} + m] \subset \mathcal{M}[u;h]$ for all $m > 0$, i.e. by (123), that

$$h'(u)\psi^2 h_1(\widetilde{u}+m)h_1'(\widetilde{u}+m) \in W_0^{1,2}(\Omega)$$

whenever $\psi \in C_0^{0,1}(\Omega)$, $m > 0$, h_1 is admissible for $\widetilde{u} + m$ where $\widetilde{u} = h(u)$, and h is as above. Now fix $m > 0$ and set

$$h_2(t) = h_1(t+m),$$
$$h_3(t) = h_2(h(t)) = h_2 \circ h(t).$$

Then $h_3' = (h_2' \circ h)h'$ implies that

$$h'(u)\psi^2 h_1(\widetilde{u}+m)h_1'(\widetilde{u}+m) = h'(u)\psi^2 h_2(\widetilde{u})h_2'(\widetilde{u}) = \psi^2 h_3(u)h_3'(u),$$

and thus by (119) applied to h_3, matters are reduced to proving that h_3 is admissible on $\mathbb{R}$. Clearly $h_3 \in C^1(\mathbb{R})\cap C_{pw}^2(\mathbb{R})$ is positive and monotone, and $h_3' = (h_2' \circ h)h'$ is bounded. To establish the remaining inequalities in (108) for h_3, we compute

$$h_3'' = (h_2'' \circ h)(h')^2 + (h_2' \circ h)h''.$$

Clearly, h_3'' is bounded and

$$
\begin{aligned}
th_3''(t) &= th_2''(h(t))(h'(t))^2 + th_2'(h(t))h''(t) \\
&= \{h(t)h_2''(h(t))\}\frac{t(h'(t))^2}{h(t)} + \{th''(t)\}h_2'(h(t)).
\end{aligned}
$$

By (108), both terms in braces are bounded, as is $h_2'(h(t))$. Finally, $\frac{t(h'(t))^2}{h(t)}$ is bounded for the special choice of $h(t)$ made above. This completes the proof that (147), with $\gamma > \frac{1}{2}\max\{\mu,\nu\}$, does indeed hold for $\widetilde{u} = h(u)$ when u is a weak subsolution of (99) in Ω.

We obtain from (147) with $\gamma > \frac{1}{2}\max\{\mu,\nu\}$ applied to $\widetilde{u}$ and equation (149) that

$$
(151)\qquad \|u_+\|_{L^\infty(B(y,cr))} \le \left\|\overline{\widetilde{u}}\right\|_{L^\infty(B(y,cr))} \le C_{\sigma,\gamma,\tau}^{\frac{1}{\gamma}}\left\{\frac{1}{|B(y,r)|}\int_{B(y,r)}\overline{\widetilde{u}}^{2\gamma}\right\}^{\frac{1}{2\gamma}},
$$

where $\overline{\widetilde{u}} = \widetilde{u} + \widetilde{m}(r)$ and

$$
\widetilde{m}(r) = r^{2\eta}\left\|\widetilde{f}\right\|_{\frac{q}{2}} + r^\eta\|\widetilde{\mathbf{g}}\|_q.
$$

We wish to show that (151) holds with $\gamma = 1$, possibly with a bigger constant and an enlarged ball on the right side. In case $\mu \ge \upsilon$, this is true with no change on the right side since the requirement that $\gamma > \frac{1}{2}\max\{\mu,\nu\}$ is satisfied for $\gamma = 1$ due to our hypothesis that $p > 4$, which implies that $\max\{\mu,\nu\} = \mu = \left(\frac{p}{2}\right)' < 2$. If on the other hand $\nu > \mu$, we recall the case $j = 1$ of (142), but with u there replaced by $\widetilde{u}$ and r there replaced by $\frac{r}{c}$ so that

$$
B(y,r) \subset \{\psi_1 = 1\} \subset supp\,\psi_1 = E_1 \subset B\left(y,\frac{r}{c}\right)
$$

and if $\beta > \frac{1}{2}$,

(152)

$$
\begin{aligned}
\left\{\frac{1}{|B(y,r)|}\int_{B(y,r)}\overline{\widetilde{u}}^{2\beta\nu}\right\}^{\frac{1}{2\nu}} &\le C\left\{\frac{1}{|E_1|}\int\left(\psi_1\overline{\widetilde{u}}^\beta\right)^{2\nu}\right\}^{\frac{1}{2\nu}} \\
&\le C\mathcal{C}(p,q,\sigma,\beta,1)\left\{\frac{1}{|E_1|}\int_{E_1}\overline{\widetilde{u}}^{2\beta\mu}\right\}^{\frac{1}{2\mu}} \\
&\le C'\mathcal{C}(p,q,\sigma,\beta,1)\left\{\frac{1}{|B(y,\frac{r}{c})|}\int_{B(y,\frac{r}{c})}\overline{\widetilde{u}}^{2\beta\mu}\right\}^{\frac{1}{2\mu}},
\end{aligned}
$$

provided $\int\left(\psi_1\overline{\widetilde{u}}^\beta\right)^{2\nu} < \infty$, but where now $\overline{\widetilde{u}} = \widetilde{u} + \widetilde{m}\left(\frac{r}{c}\right)$. Choose $\beta = \frac{\gamma}{\max\{\mu,\nu\}}$ which is a legitimate choice for β since $\frac{\gamma}{\max\{\mu,\nu\}} > \frac{1}{2}$. Then $2\beta\nu = 2\frac{\gamma}{\max\{\mu,\nu\}}\nu = 2\gamma$ since we now have $\nu > \mu$. We next claim that

- $\beta\mu \le 1$,
- $\int\left(\psi_1\overline{\widetilde{u}}^\beta\right)^{2\nu} < \infty$.

To obtain these two inequalities, we first observe that $p > 4$ implies $\mu = \left(\frac{p}{2}\right)' < 2$, and since $\max\{\mu, \nu\} = \nu$, we may choose γ such that

$$\frac{\nu}{2} < \gamma \leq \frac{\nu}{\mu}.$$

Note that since $q > 2\sigma'$, we have $\nu < \sigma$ and then

$$\gamma \leq \frac{\nu}{\mu} < \frac{\sigma}{\mu} \leq \sigma.$$

Thus $\beta\mu = \frac{\gamma\mu}{\nu} \leq 1$ and

$$\left\{\int \left(\psi_1 \overline{\overline{u}}^\beta\right)^{2\nu}\right\}^{\frac{1}{2\beta\nu}} \leq \left\{\int_{B\left(y, \frac{r}{c}\right)} \overline{\overline{u}}^{2\sigma}\right\}^{\frac{1}{2\sigma}} < \infty$$

by the Sobolev inequality (15) with $w = \psi\overline{u}$, where $\psi \in C_0^{0,1}(\Omega)$ and $\psi = 1$ on $B\left(y, \frac{r}{c}\right)$. This establishes that (152) holds with $2\beta\nu = 2\gamma$ and $2\beta\mu \leq 2$, and hence that

$$(153) \quad \left\{\frac{1}{|B(y,r)|}\int_{B(y,r)} \overline{\overline{u}}^{2\gamma}\right\}^{\frac{1}{2\gamma}} \leq C'\mathcal{C}(p, q, \sigma, \beta, 1) \left\{\frac{1}{|B\left(y, \frac{r}{c}\right)|}\int_{B\left(y, \frac{r}{c}\right)} \overline{\overline{u}}^2\right\}^{\frac{1}{2}}.$$

From (153) and the inequality $\widetilde{m}\left(\frac{r}{c}\right) \leq C\widetilde{m}(r)$ we obtain that (151) holds with $\gamma = 1$, possibly with a bigger constant and the enlarged ball $B\left(y, \frac{r}{c}\right)$ on the right side. Using the inequality

$$\overline{\overline{u}} = \widetilde{u} + r^{2\eta}\left\|\widetilde{\widetilde{f}}\right\|_{\frac{q}{2}} + r^\eta \|\widetilde{\mathbf{g}}\|_q$$

$$\leq \sqrt{m(r)^2 + u_+^2} + r^{2\eta}\|f\|_{\frac{q}{2}} + Cm(r)^{-1}\left(r^\eta \|\mathbf{g}\|_q\right)^2 + r^\eta \|\mathbf{g}\|_q$$

$$\leq u_+ + (C+2)m(r),$$

we finally obtain that

$$\|u_+\|_{L^\infty(B(y,cr))} \leq CC_{\sigma,\tau}^{\frac{1}{\gamma}} \left(\left\{\frac{1}{|B\left(y, \frac{r}{c}\right)|}\int_{B\left(y, \frac{r}{c}\right)} u_+^2\right\}^{\frac{1}{2}} + m(r)\right),$$

when u is a weak subsolution to (101). If u is a weak supersolution to $\mathcal{L}u = f + \mathbf{T}'\mathbf{g}$, then we consider the subsolution $-u$ to $\mathcal{L}u = -f - \mathbf{T}'\mathbf{g}$. This completes the proof of the local estimate on balls in Proposition 59.

3. The strong Harnack inequality

Let u be a nonnegative weak solution to (99). Then the first part of (146) holds if $\gamma \neq 0$, and thus (147) holds if $\gamma \neq 0$. For $\gamma > 0$, we multiply the following two inequalities from (147),

$$\operatorname*{ess\,sup}_{x \in B(y,cr)} \overline{u}(x) \leq C_{\sigma,\gamma,\tau}^{\frac{1}{\gamma}} \left\{\frac{1}{|B(y,r)|}\int_{B(y,r)} \overline{u}^{2\gamma}\right\}^{\frac{1}{2\gamma}}$$

and

$$\operatorname*{ess\,sup}_{x \in B(y,cr)} \overline{u}(x)^{-1} \leq C_{\sigma,-\gamma,\tau}^{\frac{1}{\gamma}} \left\{\frac{1}{|B(y,r)|}\int_{B(y,r)} \overline{u}^{-2\gamma}\right\}^{\frac{1}{2\gamma}},$$

to obtain

$$(154) \qquad \frac{\operatorname{ess\,sup}_{x \in B(y,cr)} \overline{u}\,(x)}{\operatorname{ess\,inf}_{x \in B(y,cr)} \overline{u}\,(x)}$$

$$\leq \; C_{\sigma,\gamma,\tau}^{\frac{1}{\gamma}} C_{\sigma,-\gamma,\tau}^{\frac{1}{\gamma}} \left\{ \left[\frac{1}{|B\,(y,r)|} \int_{B(y,r)} (\overline{u}^{\gamma})^2 \right] \left[\frac{1}{|B\,(y,r)|} \int_{B(y,r)} \left(\frac{1}{\overline{u}^{\gamma}} \right)^2 \right] \right\}^{\frac{1}{2\gamma}} .$$

Note that the right side of (154) is finite since $\overline{u}$ is bounded below by construction, and bounded above by Proposition 59. To obtain the "standard strong form" (172) of Harnack's inequality from (154), it only remains to show that the right side of (154) is bounded independent of the ball $B\,(y,r) \subset \Omega$ for some $\gamma > 0$, i.e. that whenever u is a nonnegative weak solution of (99) in Ω, then the weight $\overline{u}\,(x)^{2\gamma}$ is an A_2 weight for some $\gamma > 0$. By an extension of the John-Nirenberg inequality to homogeneous spaces in Stromberg and Torchinsky [43], which we recall in the short subsection below, this will follow if we can show that $v\,(x) \equiv \log \overline{u}\,(x) \in BMO$ relative to the homogeneous space with quasimetric $d\,(x,y)$, i.e. if

$$\frac{1}{|B\,(y,r)|} \int_{B(y,r)} \left(v - \frac{1}{|B\,(y,r)|} \int_{B(y,r)} v \right)^2 \leq C,$$

for all balls $B\,(y,r) \subset \Omega$. Using the Poincaré inequality (17), it then suffices by (27), which relates the norms $\| \cdot \|_Q$ and $\| \cdot \|$, to show that

$$(155) \qquad \int_{B(y,C_0 r)} \|\nabla v\|^2 \leq C' r^{-2} |B\,(y,C_0 r)|,$$

for all balls $B\,(y,C_0 r) \subset \Omega$. It will then follow that $\|v\|_{BMO} \leq \sqrt{CC'}$. However, we must localize these standard results to balls $B\,(y,r)$ with $y \in \Omega$, $0 < r < \delta \; dist\,(y,\partial\Omega)$, since we are only able to obtain (155) for such δ-local balls. See the appendix for a short discussion of an alternate method due to E. Bombieri for proving Harnack's inequality.

3.1. The John-Nirenberg inequality in a homogeneous space. Our purpose here is to recall the classical distribution inequality of F. John and L. Nirenberg [23] in the setting of homogeneous spaces, and obtain as a consequence that exponentials of BMO functions with small norm are A_2 weights. Moreover, we must obtain this in the local sense for δ-local balls $B\,(y,r)$, i.e. those with $y \in \Omega$ and $0 < r < \delta \; dist\,(y,\partial\Omega)$. Suppose Ω is an open subset of $\mathbb{R}^n$ and $(\Omega, d\,(x,y), d\mu)$ is a symmetric general homogeneous space. Specifically, this means that d is a symmetric quasimetric,

$$\begin{aligned} d\,(x,y) &= 0 \Longleftrightarrow x = y, \\ d\,(x,y) &= d\,(y,x), \\ d\,(x,y) &\leq \kappa\,(d\,(x,z) + d\,(z,y)), \end{aligned}$$

$x,y,z \in \Omega$, and the balls $B\,(x,r) = \{y \in \Omega : d\,(x,y) < r\}$ are μ-measurable and satisfy a doubling condition relative to the measure $d\mu$,

$$|B\,(x,2r)|_{\mu} \leq C_{doub} \,|B\,(x,r)|_{\mu} .$$

For a locally integrable function f on Ω, we say that $f \in BMO = BMO\,(\Omega, d, \mu)$ if

$$(156) \qquad \frac{1}{|B|_\mu} \int_B |f - f_B| \, d\mu \leq C,$$

for all balls B, where $f_B = \frac{1}{|B|_\mu} \int_B f d\mu$ is the μ-average of f on B. The infimum of all constants C satisfying (156) is the BMO "norm" of f, denoted $\|f\|_{BMO}$. We also consider the δ-local space δ- $BMO = \delta$- $BMO\,(\Omega, d, \mu)$ for $\delta > 0$ with the same definition as above, but with the balls $B = B\,(x, r)$ restricted to be δ-local, i.e. $x \in \Omega, 0 < r < \delta \, dist\,(x, \partial\Omega)$. For a locally integrable nonnegative function w on Ω, we say that $w \in A_2 = A_2\,(\Omega, d, \mu)$ if

$$(157) \qquad \left(\frac{1}{|B|_\mu} \int_B w d\mu \right) \left(\frac{1}{|B|_\mu} \int_B w^{-1} d\mu \right) \leq C$$

for all balls B. The infimum of all constants C satisfying (157) is the A_2 "norm" of w, denoted $\|w\|_{A_2}$. Similarly, we consider δ- $A_2 = \delta$- $A_2\,(\Omega, d, \mu)$ with δ-local balls in the above definition.

LEMMA 60. *Given $\delta > 0$, there are positive constants δ_0, C_1 and c_2 such that*

$$(158) \qquad |\{x \in B_0 : |f\,(x) - f_{B_0}| > \alpha\}|_\mu \leq C_1 e^{-\frac{c_2 \alpha}{\|f\|_{\delta\text{-}BMO}}} |B_0|_\mu$$

for all $\alpha > 0$, $f \in \delta$- BMO and δ_0-local balls B_0.

In the global case $\delta = \delta_0 = \infty$, this lemma is a special case of Theorem 2 in chapter III of Stromberg and Torchinsky [43]. The lemma can also be proved in this case by adapting the original argument of John and Nirenberg [23], as executed in sections 3.6 and 3.7 of chapter IV of Stein [42], to the grid of "dyadic cubes" in a homogeneous space as constructed in Christ [4], or the authors' [41]. For the reader's convenience, we give this latter proof in the local setting following the corollary on A_2 weights.

COROLLARY 61. *Given $\delta > 0$, there are positive constants δ_0, C_1 and c_2 such that $\|e^f\|_{\delta_0\text{-}A_2} \leq \left(1 + C_1 \frac{\|f\|_{\delta\text{-}BMO}}{c_2 - \|f\|_{\delta\text{-}BMO}} \right)^2$ whenever $\|f\|_{\delta\text{-}BMO} < c_2$.*

The corollary is easily obtained from the lemma by integrating (158) to obtain

$$\frac{1}{|B_0|_\mu} \int_{B_0} e^{|f - f_{B_0}|} d\mu = \frac{1}{|B_0|_\mu} \int_{-\infty}^{\infty} e^\alpha |\{x \in B_0 : |f - f_{B_0}| > \alpha\}|_\mu \, d\alpha$$

$$\leq 1 + C_1 \int_0^\infty e^{\alpha\left(1 - \frac{c_2}{\|f\|_{\delta\text{-}BMO}}\right)} d\alpha = 1 + C_1 \frac{\|f\|_{\delta\text{-}BMO}}{c_2 - \|f\|_{\delta\text{-}BMO}}$$

for δ_0-local balls B_0 whenever $\|f\|_{\delta\text{-}BMO} < c_2$, and then computing

$$\left(\frac{1}{|B_0|_\mu} \int_{B_0} e^f d\mu\right) \left(\frac{1}{|B_0|_\mu} \int_{B_0} e^{-f} d\mu\right)$$

$$= \left(\frac{1}{|B_0|_\mu} \int_{B_0} e^{(f-f_{B_0})} d\mu\right) \left(\frac{1}{|B_0|_\mu} \int_{B_0} e^{-(f-f_{B_0})} d\mu\right)$$

$$\leq \left(\frac{1}{|B_0|_\mu} \int_{B_0} e^{|f-f_{B_0}|} d\mu\right) \left(\frac{1}{|B_0|_\mu} \int_{B_0} e^{|f-f_{B_0}|} d\mu\right)$$

$$\leq \left(1 + C_1 \frac{\|f\|_{\delta\text{-}BMO}}{c_2 - \|f\|_{\delta\text{-}BMO}}\right)^2.$$

PROOF (OF LEMMA 60). We begin by recalling the grid of "dyadic cubes" in a homogeneous space. There is a constant $\lambda > 1$ such that for every $m \in \mathbb{Z}$, there are points $x_j^k \in \Omega$ and Borel sets Q_j^k, $1 \leq j < n_k$, $k \geq m$ (where $n_k \in \mathbb{N} \cup \{\infty\}$) depending on m, such that
(159)
$$\begin{cases} B\left(x_j^k, \lambda^k\right) \subset Q_j^k \subset B\left(x_j^k, \lambda^{k+1}\right) & for \quad 1 \leq j < n_k, k \geq m \\ \Omega = \cup_j Q_j^k, & for \quad k \geq m \\ Q_i^k \cap Q_{i'}^k = \phi & for \quad k \geq m, i \neq i' \\ Either \ Q_j^k \subset Q_i^\ell \ or \ Q_j^k \cap Q_i^\ell = \phi & for \quad 1 \leq j < n_k, 1 \leq i < n_\ell, m \leq k < \ell \end{cases}$$

We denote by $\mathcal{D}_m$ the collection of the "dyadic cubes" Q_j^k for $k \geq m$. For a construction of such a collection $\mathcal{D}$ in the case $m = -\infty$, see M. Christ [4]. A construction of the collection $\mathcal{D}_m$ for $m \in \mathbb{Z}$ appears in the authors' paper [41]. We say that a dyadic cube Q_j^k is δ-local if the ball $B\left(x_j^k, \lambda^{k+1}\right)$ is δ-local, i.e. $0 < \lambda^{k+1} < \delta \ dist\left(x_j^k, \partial\Omega\right)$, and we write δ-$\mathcal{D}_m$ for the collection of δ-local dyadic cubes Q_j^k for $k \geq m$.

Fix $m \in \mathbb{Z}$. We begin by establishing the dyadic distribution inequality: there exist C_1, $c_2 > 0$ independent of m such that

(160) $$|\{x \in Q_0 : |f^m(x) - f_{Q_0}| > \alpha\}|_\mu \leq C_1 e^{-\frac{c_2\alpha}{\|f\|_{\delta\text{-}BMO}}} |Q_0|_\mu$$

for all $\alpha > 0$, dyadic cubes $Q_0 \in \delta$-$\mathcal{D}_m$, and $f \in \delta$-BMO with $\|f\|_{\delta\text{-}BMO} = 1$. The function f^m is the expectation of f on the dyadic decomposition $\Omega = \{Q_j^m\}_j$:

(161) $$f^m(x) = \sum_j f_{Q_j^m} \chi_{Q_j^m}(x) = \sum_j \left(\frac{1}{|Q_j^m|_\mu} \int_{Q_j^m} f d\mu\right) \chi_{Q_j^m}(x).$$

Let M_μ^Δ denote the dyadic maximal operator

$$M_\mu^\Delta h(x) = \sup_{x \in Q} \frac{1}{|Q|_\mu} \int_Q |h| d\mu,$$

where it is understood that the supremum is taken over all dyadic cubes $Q \in \delta\text{-}\mathcal{D}_m$. Now if $Q_j^k \in \delta\text{-}\mathcal{D}_m$, then by (159),

$$\left| f_{B\left(x_j^k, \lambda^{k+1}\right)} - f_{Q_j^k} \right| = \left| \frac{1}{\left|Q_j^k\right|_\mu} \int_{Q_j^k} \left(f - f_{B\left(x_j^k, \lambda^{k+1}\right)} \right) d\mu \right|$$

$$\leq \frac{1}{\left|B\left(x_j^k, \lambda^k\right)\right|_\mu} \int_{B\left(x_j^k, \lambda^{k+1}\right)} \left| f - f_{B\left(x_j^k, \lambda^{k+1}\right)} \right| d\mu.$$

We then have for $Q = Q_j^k \in \delta\text{-}\mathcal{D}_m$,

$$(162) \qquad \frac{1}{|Q|_\mu} \int_Q |f - f_Q| \, d\mu$$

$$\leq \frac{1}{|Q|_\mu} \int_Q \left| f - f_{B\left(x_j^k, \lambda^{k+1}\right)} \right| d\mu + \left| f_{B\left(x_j^k, \lambda^{k+1}\right)} - f_Q \right|$$

$$\leq \frac{2}{\left|B\left(x_j^k, \lambda^k\right)\right|_\mu} \int_{B\left(x_j^k, \lambda^{k+1}\right)} \left| f - f_{B\left(x_j^k, \lambda^{k+1}\right)} \right| d\mu$$

$$\leq \frac{2C_0}{\left|B\left(x_j^k, \lambda^{k+1}\right)\right|_\mu} \int_{B\left(x_j^k, \lambda^{k+1}\right)} \left| f - f_{B\left(x_j^k, \lambda^{k+1}\right)} \right| d\mu \leq 2C_0,$$

since $\|f\|_{\delta\text{-}BMO} = 1$, and where C_0 is a doubling constant such that $|B(x, r\lambda)|_\mu \leq C_0 \, |B(x, r)|_\mu$.

Fix a dyadic cube $Q_0 \in \delta\text{-}\mathcal{D}_m$ and let $h = (f - f_{Q_0}) \chi_{Q_0}$. From (162) we obtain that for $Q \supset Q_0$,

$$\frac{1}{|Q|_\mu} \int_Q |h| \, d\mu \leq \frac{1}{|Q_0|_\mu} \int_{Q_0} |f - f_{Q_0}| \, d\mu \leq 2C_0.$$

Now for $\alpha > 0$, let

$$\Omega_\alpha = \left\{ x \in \Omega : M_\mu^\triangle h(x) > \alpha \right\},$$

and note that for $\alpha \geq 2C_0$, the inequality above, plus the dyadic structure, yields $\Omega_\alpha \subset Q_0$. Let $\mathcal{C}_\alpha$ be the collection of dyadic cubes $Q \in \delta\text{-}\mathcal{D}_m$ such that the average of $|h|$ over Q exceeds α. Note that the cubes in $\mathcal{C}_\alpha$ are proper subcubes of Q_0. Then let $\{Q_{\alpha,j}\}_j$ be the collection of maximal dyadic cubes in $\mathcal{C}_\alpha$. Then we have

   (a) for each $\alpha \geq 2C_0$, the $Q_{\alpha,j}$ are pairwise disjoint,
   (b) for each $\alpha \geq 2C_0$, $\cup_j Q_{\alpha,j} = \Omega_\alpha \subset Q_0$,
   (c) If $2C_0 \leq \alpha < \beta$, and i, j are given, then either $Q_{\beta,j} \subset Q_{\alpha,i}$ or $Q_{\beta,j} \cap Q_{\alpha,i} = \phi$.

To see part (c), note that if the cubes intersect, then one is contained in the other by the dyadic structure, and so $Q_{\beta,j} \supset Q_{\alpha,i}$ properly would violate maximality. Now denote the distribution function of $M_\mu^\triangle h(x)$ by

$$\lambda(\alpha) = |\Omega_\alpha|_\mu = \left| \left\{ x \in \Omega : M_\mu^\triangle h(x) > \alpha \right\} \right|_\mu.$$

Let C_1 be a doubling constant for the dyadic cubes:

$$(163) \qquad \left| \widetilde{Q} \right|_\mu \leq \left| B\left(x_i^{k+1}, \lambda^{k+2}\right) \right|_\mu \leq C_1 \left| B\left(x_j^k, \lambda^k\right) \right|_\mu \leq C_1 |Q|_\mu,$$

whenever $Q = Q_j^k \supset B\left(x_j^k, \lambda^k\right)$ and $\widetilde{Q} = Q_i^{k+1} \subset B\left(x_i^{k+1}, \lambda^{k+2}\right)$ with $Q \subset \widetilde{Q}$. We claim that with $\gamma = \gamma(\alpha) = 1 + \frac{4C_0C_1}{\alpha}$ and $\alpha \geq 2C_0$,

$$(164) \qquad |\Omega_{\gamma\alpha} \cap Q_{\alpha,j}|_\mu \leq \frac{1}{2}|Q_{\alpha,j}|_\mu, \qquad \text{for all } j.$$

To see this, note first that if $\widetilde{Q_{\alpha,j}}$ is the dyadic predecessor of $Q_{\alpha,j}$, then $\widetilde{Q_{\alpha,j}} \subset Q_0$ and

$$\left|h_{\widetilde{Q_{\alpha,j}}}\right| \leq \frac{1}{\left|\widetilde{Q_{\alpha,j}}\right|_\mu}\int_{\widetilde{Q_{\alpha,j}}}|h|\,d\mu \leq \alpha,$$

by the maximality of the dyadic cubes $Q_{\alpha,j}$. Now let $g = \chi_{\widetilde{Q_{\alpha,j}}}\left(h - h_{\widetilde{Q_{\alpha,j}}}\right)$. Take $\gamma \geq 1$ and $x \in \Omega_{\gamma\alpha} \cap Q_{\alpha,j}$. By properties (b) and (c) above, there is, for some i, a dyadic cube $Q_{\gamma\alpha,i}$ containing x with $Q_{\gamma\alpha,i} \subset Q_{\alpha,j}$ such that

$$\gamma\alpha < \frac{1}{|Q_{\gamma\alpha,i}|_\mu}\int_{Q_{\gamma\alpha,i}}|h|\,d\mu \leq \frac{1}{|Q_{\gamma\alpha,i}|_\mu}\int_{Q_{\gamma\alpha,i}}|g|\,d\mu + \left|h_{\widetilde{Q_{\alpha,j}}}\right| \leq \frac{1}{|Q_{\gamma\alpha,i}|_\mu}\int_{Q_{\gamma\alpha,i}}|g|\,d\mu + \alpha.$$

Thus we have

$$M_\mu^\Delta g(x) > \gamma\alpha - \alpha = (\gamma - 1)\alpha, \qquad \text{for } x \in \Omega_{\gamma\alpha} \cap Q_{\alpha,j}.$$

Now we use the nested property of dyadic cubes in a crucial way. Since $\widetilde{Q_{\alpha,j}} \subset Q_0$, we have for $x \in \widetilde{Q_{\alpha,j}}$,

$$g(x) = h(x) - h_{\widetilde{Q_{\alpha,j}}} = (f(x) - f_{Q_0}) - (f - f_{Q_0})_{\widetilde{Q_{\alpha,j}}} = f(x) - f_{\widetilde{Q_{\alpha,j}}},$$

and thus the weak type inequality for the dyadic maximal operator M_μ^Δ (which has constant 1) yields

$$\begin{aligned}
|\Omega_{\gamma\alpha} \cap Q_{\alpha,j}|_\mu &\leq \left|\left\{M_\mu^\Delta g > (\gamma - 1)\alpha\right\}\right|_\mu \leq \frac{1}{(\gamma-1)\alpha}\int|g|\,d\mu \\
&= \frac{1}{(\gamma-1)\alpha}\int_{\widetilde{Q_{\alpha,j}}}\left|f - f_{\widetilde{Q_{\alpha,j}}}\right|\,d\mu \leq \frac{1}{(\gamma-1)\alpha}2C_0\left|\widetilde{Q_{\alpha,j}}\right|_\mu \\
&\leq \frac{1}{(\gamma-1)\alpha}2C_0C_1|Q_{\alpha,j}|_\mu,
\end{aligned}$$

for $\alpha \geq 2C_0$ by (162) and (163). This establishes (164) with $\gamma = \gamma(\alpha) = 1 + \frac{4C_0C_1}{\alpha}$. Now sum (164) in j to obtain

$$(165) \qquad \lambda(\alpha + 4C_0C_1) = \lambda(\gamma(\alpha)\alpha) = |\Omega_{\gamma\alpha}|_\mu \leq \frac{1}{2}|\Omega_\alpha|_\mu = \frac{1}{2}\lambda(\alpha),$$

for all $\alpha \geq 2C_0$. We now obtain that with $c_2 = \frac{\log 2}{4C_0C_1}$,

$$\left|\left\{x \in Q_0 : M_\mu^\Delta h(x) > \alpha\right\}\right|_\mu \leq C'e^{-c_2\alpha}|Q_0|_\mu$$

for all $\alpha \geq 2C_0$ by iterating (165) and using $|\Omega_{2C_0}|_\mu \leq |Q_0|_\mu$. For $\alpha \leq 2C_0$, we simply use $\left\{x \in Q_0 : M_\mu^\Delta h(x) > \alpha\right\} \subset Q_0$, and increase C' to $e^{2C_0c_2}$ if necessary, to obtain that

$$\left|\left\{x \in Q_0 : M_\mu^\Delta\left[(f - f_{Q_0})\chi_{Q_0}\right](x) > \alpha\right\}\right|_\mu \leq C'e^{-c_2\alpha}|Q_0|_\mu$$

for all $\alpha > 0$. Next we observe that

$$M_\mu^\Delta\left[(f - f_{Q_0})\chi_{Q_0}\right] \geq |f^m - f_{Q_0}|\chi_{Q_0}$$

by the dyadic structure (159), and this proves (160).

We can now obtain (158) as follows. There is a positive constant C with the following property: Given a ball $B_0 = B(x_0, r)$ and $m \in \mathbb{Z}$ with $\lambda^m \ll r$, let $k > m$ be determined by $\lambda^k < r \leq \lambda^{k+1}$. Then there exist dyadic cubes $\{Q_j^k\}_{j \in F}$ in $\mathcal{D}_m$ such that

$$B_0 \subset \cup_{j \in F} Q_j^k \subset B_0^* \equiv B(x_0, Cr).$$

A packing argument shows that $\#F \leq C$. Indeed, the engulfing and doubling properties of the balls show there are positive constants C', C'' such that

$$|B_0^*|_\mu = |B(x_0, Cr)|_\mu \leq |B(x_j^k, C''\lambda^k)|_\mu \leq C' |B(x_j^k, \lambda^k)|_\mu \leq C' |Q_j^k|_\mu$$

for $j \in F$, and adding these inequalities we obtain

$$(\#F)|B_0^*|_\mu \leq C' \sum_{j \in F} |Q_j^k|_\mu \leq C |B_0^*|_\mu.$$

Now choose $\delta_0 > 0$ sufficiently small that the ball $B(x, Cr)$ is δ-local whenever $x \in \Omega$ and the ball $B(x, r)$ is δ_0-local. Then if $B_0 = B(x_0, r)$ is δ_0-local, and E is any subset of B_0^* with $|E|_\mu \geq c|B_0^*|_\mu$, we have

$$
\begin{aligned}
|f_{B_0^*} - f_E| &= \left| \frac{1}{|E|_\mu} \int_E (f - f_{B_0^*}) \, d\mu \right| \\
&\leq \frac{C}{|B_0^*|_\mu} \int_{B_0^*} |f - f_{B_0^*}| \, d\mu \leq C
\end{aligned}
$$

since $\|f\|_{\delta\text{-}BMO} = 1$ and B_0^* is δ-local. It follows with $E = B_0, Q_j^k$ that

$$(166) \qquad \left| f_{B_0} - f_{Q_j^k} \right| \leq 2C, \quad j \in F.$$

Thus for $\alpha > 4C$, (166) shows that $\left| f^m(x) - f_{Q_j^k} \right| > \frac{\alpha}{2}$ whenever $|f^m(x) - f_{B_0}| > \alpha$, and so by (160),

$$
\begin{aligned}
(167) \quad |\{x \in B_0 : |f^m(x) - f_{B_0}| > \alpha\}|_\mu &\leq \sum_{j \in F} \left| \left\{ x \in Q_j^k : \left| f^m(x) - f_{Q_j^k} \right| > \frac{\alpha}{2} \right\} \right|_\mu \\
&\leq \sum_{j \in F} C' e^{-c_2 \frac{\alpha}{2}} |Q_j^k|_\mu \\
&\leq C'' e^{-\frac{c_2}{2}\alpha} |B_0|_\mu,
\end{aligned}
$$

and hence also for all $\alpha > 0$ upon increasing C'' to $e^{\frac{c_2}{2}4C}$ if necessary.

We have that $f^m(x) \to f(x)$ for μ-almost every x in B_0 by a standard generalization of the corollary concerning differentiation of integrals on page 13 of [42]. In establishing this generalization, we use the fact that our standing assumption (74), which is equivalent to the hypothesis $d(x, y) \geq c|x - y|$ of Theorem 8 with $c = C_{euc}^{-1}$, shows that the Euclidean diameters of the balls $B(x, r)$ shrink to 0 as $r \to 0$. Thus the Euclidean diameters of the dyadic cubes Q_j^m also shrink to 0 as $m \to -\infty$, and this yields that $\lim_{m \to -\infty} \frac{1}{|Q_j^m|} \int_{Q_j^m} g = g(x)$ if g is continuous at x, where it is understood that $x \in Q_j^m$. The maximal theorem on page 13 of [42] then completes the proof that $\lim_{m \to -\infty} f^m = f$ μ-a.e. using a familiar argument.

We thus obtain from Fatou's lemma, upon letting $m \to -\infty$ in (167), that

$$
\begin{aligned}
|\{x \in B_0 : |f(x) - f_{B_0}| > \alpha\}|_\mu &= \int \chi_{\{x \in B_0 : |f(x) - f_{B_0}| > \alpha\}} d\mu \\
&\leq \int \liminf_{m \to -\infty} \chi_{\{x \in B_0 : |f^m(x) - f_{B_0}| > \alpha\}} d\mu \\
&\leq \liminf_{m \to -\infty} \int \chi_{\{x \in B_0 : |f^m(x) - f_{B_0}| > \alpha\}} d\mu \\
&= \liminf_{m \to -\infty} |\{x \in B_0 : |f^m(x) - f_{B_0}| > \alpha\}|_\mu \\
&\leq C'' e^{-\frac{c_2}{2}\alpha} |B_0|_\mu,
\end{aligned}
$$

for all $\alpha > 0$. This proves (158) for $\|f\|_{\delta\text{-}BMO} = 1$, and the general case follows upon replacing f and α with $\dfrac{f}{\|f\|_{\delta\text{-}BMO}}$ and $\dfrac{\alpha}{\|f\|_{\delta\text{-}BMO}}$.

3.2. Logarithmic energy inequality for positive solutions. Here we will obtain (155) for a δ-local ball $B(y, C_0 r)$ with $v = \log \overline{u}$ for a nonnegative solution u to (99), in fact for u a nonnegative supersolution to (99). To see this, let y and r be as above, i.e. $y \in \Omega$ and $0 < C_0 r < \delta \, dist(y, \partial\Omega)$, and let

$$
\mathcal{H}(t) = \log \frac{1}{t + m(r)}, \qquad \text{for } t > 0,
$$

with $m(r) = r^{2\eta} \|f\|_{\frac{q}{2}} + r^\eta \|\mathbf{g}\|_q$ and $0 < \eta < 1 - \frac{Q}{q}$ as above, so that $v(x) = \log \overline{u}(x) = -\mathcal{H}(u(x))$. As usual, if $m(r) = 0$, we replace it with $m > 0$ and later let $m \to 0$. We shall not use equation (116) for v here. Instead, we use equation (99) for u along with the standard choice of testing function $w = -\psi^2 \mathcal{H}'(u) = \psi^2 (u + m(r))^{-1}$ (see however Remark 57, where this choice arises as $w = \psi^2 h(u) h'(u)$ in the class $\mathcal{A}[u]$ with $h(t) = \sqrt{2 \log \frac{t + m(r)}{m(r)}}$). Note the equalities

$$
\begin{aligned}
(168) \qquad \mathcal{H}'(t) &= -(t + m(r))^{-1}, \\
\mathcal{H}''(t) &= (t + m(r))^{-2}, \\
|\mathcal{H}'(t)|^2 &= \mathcal{H}''(t),
\end{aligned}
$$

for $0 < t < \infty$. Let $\psi = \psi_1$ be as in (20). With $w = -\psi^2 \mathcal{H}'(u)$ in (102), we obtain

$$
\begin{aligned}
(169) \quad \int \psi^2 \mathcal{H}''(u) \|\nabla u\|^2 \\
\leq -2 \int \langle \psi \mathcal{H}'(u) \nabla u, \nabla \psi \rangle - \int \psi^2 \mathcal{H}'(u) f + \int \psi^2 \mathcal{H}'(u) uF \\
-2 \int \psi \mathcal{H}'(u) \mathbf{g} \mathbf{T} \psi - \int \psi^2 \mathcal{H}''(u) \mathbf{g} \mathbf{T} u + 2 \int \psi \mathcal{H}'(u) u \mathbf{G} \mathbf{S} \psi \\
+ \int \psi^2 \mathcal{H}''(u) u \mathbf{G} \mathbf{S} u + \int \psi^2 \mathcal{H}'(u) \mathbf{H} \mathbf{R} u.
\end{aligned}
$$

Applying Cauchy's inequality to the first integral on the right side of (169), we have for any $0 < \varepsilon < 1$,

$$
\begin{aligned}
\left| 2 \int \langle \psi \mathcal{H}'(u) \nabla u, \nabla \psi \rangle \right| &\leq \varepsilon^2 \int \psi^2 \mathcal{H}'(u)^2 \|\nabla u\|^2 + \frac{1}{\varepsilon^2} \int \|\nabla \psi\|^2 \\
&\leq \varepsilon^2 \int \psi^2 \mathcal{H}'(u)^2 \|\nabla u\|^2 + C \frac{1}{\varepsilon^2} r^{-2} |B(y, r)|,
\end{aligned}
$$

since Hölder's inequality, $p \geq 2$, (27) and (20) yield

$$
\begin{aligned}
\int \|\nabla \psi\|^2 &\leq |B(y,r)| \left\{ \frac{1}{|B(y,r)|} \int_{B(y,r)} \|\nabla \psi\|^p \right\}^{\frac{2}{p}} \\
&\leq C_{sym} C_p^2 r^{-2} |B(y,r)|.
\end{aligned}
$$

Applying Cauchy's inequality to the fifth integral on the right side of (169), we have

$$
\begin{aligned}
\left| \int \psi^2 \mathcal{H}''(u) \mathbf{g} \mathbf{T} u \right| &\leq \varepsilon^2 \int \psi^2 \mathcal{H}''(u) |\mathbf{T} u|^2 + \frac{1}{\varepsilon^2} \int \psi^2 \mathcal{H}''(u) |\mathbf{g}|^2 \\
&\leq C\varepsilon^2 \int \psi^2 \mathcal{H}''(u) \|\nabla u\|^2 + C \frac{1}{\varepsilon^2} m(r)^{-2} \|\mathbf{g}\|_q^2 |B(y,r)|^{1-\frac{2}{q}} \\
&\leq C\varepsilon^2 \int \psi^2 \mathcal{H}''(u) \|\nabla u\|^2 + C \frac{1}{\varepsilon^2} r^{-2\eta} |B(y,r)|^{1-\frac{2}{q}} \\
&\leq C\varepsilon^2 \int \psi^2 \mathcal{H}''(u) \|\nabla u\|^2 + C \frac{1}{\varepsilon^2} r^{-2} |B(y,r)|,
\end{aligned}
$$

by (168), since the vector fields $\mathbf{T}$ are subunit, and since $q(1-\eta) > Q$ implies $|B(y,r)| \geq cr^{q(1-\eta)}$ by (10). Now $\mathcal{H}''(u) u^2$ is bounded by (168), and so we estimate the seventh integral on the right side of (169) similarly by

$$
\begin{aligned}
\left| \int \psi^2 \mathcal{H}''(u) u \mathbf{G} \mathbf{S} u \right| &\leq \varepsilon^2 \int \psi^2 \mathcal{H}''(u) |\mathbf{S} u|^2 + \frac{1}{\varepsilon^2} \int \psi^2 \mathcal{H}''(u) u^2 |\mathbf{G}|^2 \\
&\leq C\varepsilon^2 \int \psi^2 \mathcal{H}''(u) \|\nabla u\|^2 + C \frac{1}{\varepsilon^2} \|\mathbf{G}\|_q^2 |B(y,r)|^{1-\frac{2}{q}} \\
&\leq C\varepsilon^2 \int \psi^2 \mathcal{H}''(u) \|\nabla u\|^2 + C \frac{1}{\varepsilon^2} \|\mathbf{G}\|_q r^{-2} |B(y,r)|,
\end{aligned}
$$

since the vector fields $\mathbf{S}$ are subunit, and since $q > Q$ implies $|B(y,r)| \geq cr^q$ by (10).

The second, third, fourth and sixth integrals on the right side of (169) are similarly estimated by

$$
\begin{aligned}
\left| \int \psi^2 \mathcal{H}'(u) f \right| &\leq C \|f\|_{\frac{q}{2}} m(r)^{-1} |B(y,r)|^{1-\frac{2}{q}} \\
&\leq Cr^{-2\eta} |B(y,r)|^{1-\frac{2}{q}} \\
&\leq Cr^{-2} |B(y,r)|,
\end{aligned}
$$

and

$$
\left| \int \psi^2 \mathcal{H}'(u) u F \right| \leq C \|F\|_{\frac{q}{2}} |B(y,r)|^{1-\frac{2}{q}} \leq C \|F\|_{\frac{q}{2}} r^{-2} |B(y,r)|,
$$

using that $\mathcal{H}'(u) u$ is bounded by (168), and also both

$$
\begin{aligned}
\left| 2 \int \psi \mathcal{H}'(u) \mathbf{g} \mathbf{T} \psi \right| &\leq C \|\mathbf{g}\|_q \|\mathbf{T}\psi\|_p m(r)^{-1} |B(y,r)|^{1-\frac{1}{q}-\frac{1}{p}} \\
&\leq Cr^{-\eta} \|\mathbf{T}\psi\|_p |B(y,r)|^{1-\frac{1}{q}-\frac{1}{p}} \\
&\leq Cr^{-2} |B(y,r)|,
\end{aligned}
$$

and

$$\left| 2 \int \psi \mathcal{H}'(u)\, u \mathbf{GS}\psi \right| \leq C \|\mathbf{G}\|_q \|\mathbf{S}\psi\|_p |B(y,r)|^{1-\frac{1}{q}-\frac{1}{p}}$$
$$\leq C \|\mathbf{G}\|_q r^{-2} |B(y,r)|,$$

since

$$\|\mathbf{T}\psi\|_p + \|\mathbf{S}\psi\|_p \leq C \|\,|\nabla\psi|\,\|_p \leq C C_{sym}^{\frac{1}{2}} C_p |B(y,r)|^{\frac{1}{p}} r^{-1}$$

by (20) and (27). Finally, the eighth term on the right side of (169) satisfies

$$\left| \int \psi^2 \mathcal{H}'(u) \mathbf{HR}u \right| \leq \varepsilon^2 \int \psi^2 \mathcal{H}'(u)^2 |\mathbf{R}u|^2 + \frac{1}{\varepsilon^2} \int \psi^2 |\mathbf{H}|^2$$
$$\leq \varepsilon^2 \int \psi^2 \mathcal{H}'(u)^2 \|\nabla u\|^2 + C \frac{1}{\varepsilon^2} \|\mathbf{H}\|_q^2 r^{-2} |B(y,r)|.$$

Combining these estimates yields

$$(170) \qquad \int \psi^2 \left[(1 - C\varepsilon^2) \mathcal{H}''(u) - \varepsilon^2 \mathcal{H}'(u)^2 \right] \|\nabla u\|^2 \leq C \frac{1}{\varepsilon^2} r^{-2} |B(y,r)|.$$

Now $\mathcal{H}'' = (\mathcal{H}')^2$ by (168), and so

$$\left[(1 - C\varepsilon^2) \mathcal{H}''(u) - \varepsilon^2 \mathcal{H}'(u)^2 \right] \|\nabla u\|^2 = \left[(1 - C\varepsilon^2 - \varepsilon^2) \mathcal{H}'(u)^2 \right] \|\nabla u\|^2$$
$$= (1 - C\varepsilon^2 - \varepsilon^2) \|\nabla v\|^2.$$

Thus with $\varepsilon > 0$ small enough, we obtain

$$(171) \qquad \int \psi^2 \|\nabla v\|^2 \leq C' r^{-2} |B(y,r)|,$$

and from (171) and the doubling assumption on the balls $B(y,r)$, we obtain (155) whenever u is a nonnegative supersolution of (99) and $v = -\mathcal{H}(u)$ with $\mathcal{H}$ as above. Note that the constants C in (170) and C' in (171) depend on $\|F\|_{\frac{q}{2}}$, $\|\mathbf{G}\|_q$ and $\|\mathbf{H}\|_q$, but not on $\|f\|_{\frac{q}{2}}$ or $\|\mathbf{g}\|_q$.

Inequalities (154) and (155), together with Corollary 61, prove the following strong Harnack inequality.

THEOREM 62. *Suppose that u is a nonnegative weak solution to (99) in $B(y,r)$ with $y \in \Omega$, $0 < r < \delta\, \mathrm{dist}(y, \partial\Omega)$ for sufficiently small $\delta > 0$, that (24), (23), (15) and (17) hold for some $q > Q = \max\{Q^*, 2\sigma'\}$, as well as (20) for some $p > \max\{2\sigma', 4\}$. Then u satisfies the strong Harnack inequality*

$$(172) \qquad ess \sup_{x \in B(y,c_2 r)} (u(x) + m_{q,\eta}(r)) \leq C_{Har}\, ess \inf_{x \in B(y,c_2 r)} (u(x) + m_{q,\eta}(r)),$$

where $m_{q,\eta}(r) = r^{2\eta} \|f\|_{\frac{q}{2}} + r^\eta \|\mathbf{g}\|_q$, $q(1-\eta) > Q = \max\{Q^*, 2\sigma'\}$, and the constants c_2 and C_{Har} depend on $\|F\|_{\frac{q}{2}}$, $\|\mathbf{G}\|_q$ and $\|\mathbf{H}\|_q$, but are independent of u, $B(y,r)$, $\|f\|_{\frac{q}{2}}$ and $\|\mathbf{g}\|_q$.

4. Hölder continuity of solutions

We can now deduce Hölder continuity from the Harnack inequality (172) in the usual way. Let $u \in W^{1,2}(\Omega)$ be a weak solution of (99) in Ω. Then u is locally bounded by Proposition 59.

Fix $y \in \Omega$ and $0 < \rho < 1$ such that $B(y, C\rho)$ is δ-local. For $0 < r \le \rho$, define the oscillation

$$\omega(r) = \underset{x \in B(y,r)}{\text{ess sup}}\, u(x) - \underset{x \in B(y,r)}{\text{ess inf}}\, u(x).$$

We may assume $\omega(r) > 0$ since otherwise there is nothing to prove. Clearly we have $\omega(r) < \infty$, and by adding the constant

$$M = -\frac{1}{2}\left(\underset{B(y,r)}{\text{ess sup}}\, u + \underset{B(y,r)}{\text{ess inf}}\, u\right)$$

to u we may assume

$$\underset{x \in B(y,r)}{\text{ess sup}}\,(u(x) + M) = -\underset{x \in B(y,r)}{\text{ess inf}}\,(u(x) + M) = \frac{1}{2}\omega(r).$$

Then $u_+ = 1 + \frac{u+M}{\frac{1}{2}\omega(r)}$ and $u_- = 1 - \frac{u+M}{\frac{1}{2}\omega(r)}$ satisfy ($u_\pm$ does not have the usual meaning here)

$$\mathcal{L}u_\pm = \pm\frac{f}{\frac{1}{2}\omega(r)} \pm \mathbf{T}'\left(\frac{\mathbf{g}}{\frac{1}{2}\omega(r)}\right) + F\left(1 \pm \frac{M}{\frac{1}{2}\omega(r)}\right) + \mathbf{S}'\mathbf{G}\left(1 \pm \frac{M}{\frac{1}{2}\omega(r)}\right),$$

they are both nonnegative in $B(y, r)$, and either $\text{ess sup}_{B(y,c_2r)}\, u_+ \ge 1$ or $\text{ess sup}_{B(y,c_2r)}\, u_- \ge 1$ since $u_+ + u_- = 2$. Suppose the former holds. Then

$$\underset{B(y,c_2r)}{\text{ess sup}}\,(u_+ + \widetilde{m}(r)) \ge \underset{B(y,c_2r)}{\text{ess sup}}\, u_+ \ge 1,$$

and the Harnack inequality (172) shows that

$$u_+(x) + \widetilde{m}(r) \ge \mathfrak{c} > 0$$

for *a.e.* $x \in B(y, c_2r)$, where $\mathfrak{c} = C_{Har}^{-1}$,

$$\begin{aligned}
\widetilde{m}(r) &= r^{2\eta}\left\|\frac{f}{\frac{1}{2}\omega(r)} + F\left(1 + \frac{M}{\frac{1}{2}\omega(r)}\right)\right\|_{\frac{q}{2}} \\
&\quad + r^\eta\left\{\left\|\frac{\mathbf{g}}{\frac{1}{2}\omega(r)}\right\|_q^q + \left\|\mathbf{G}\left(1 + \frac{M}{\frac{1}{2}\omega(r)}\right)\right\|_q^q\right\}^{\frac{1}{q}} \\
&\le 2\omega(r)^{-1}\left\{m(r) + r^{2\eta}\|F\|_{\frac{q}{2}}\left|\frac{1}{2}\omega(r) + M\right| + r^\eta\|\mathbf{G}\|_q\left|\frac{1}{2}\omega(r) + M\right|\right\} \\
&\le 2\omega(r)^{-1}\left\{m(r) + r^{2\eta}\|F\|_{\frac{q}{2}}N(r) + r^\eta\|\mathbf{G}\|_q N(r)\right\},
\end{aligned}$$

and

$$N(r) = \|\chi_{B(y,r)}u\|_{L^\infty}$$

is nondecreasing in r, and $0 < \mathfrak{c} < 1$ depends on $\|F\|_{\frac{q}{2}}$, $\|\mathbf{G}\|_q$ and $\|\mathbf{H}\|_q$, but is *independent* of $\|f\|_{\frac{q}{2}}$ and $\|\mathbf{g}\|_q$. Note that we have used the inequality

$$\left|\frac{1}{2}\omega(r) + M\right| = \left|\underset{B(y,r)}{\text{ess inf}}\, u\right| \le N(r).$$

Thus with

$$\sigma(r) = m(r) + N(r)\left(r^{2\eta}\|F\|_{\frac{q}{2}} + r^{\eta}\|\mathbf{G}\|_q\right),$$

we have σ nondecreasing and

$$\mathfrak{c} \le u_+(x) + \widetilde{m}(r) \le 1 + \frac{u(x) + M}{\frac{1}{2}\omega(r)} + 2\omega(r)^{-1}\sigma(r),$$

or

$$-\frac{1}{2}\omega(r)(1 - \mathfrak{c}) - \sigma(r) \le u(x) + M \le \frac{1}{2}\omega(r),$$

for a.e. $x \in B(y, c_2 r)$, and it follows that

(173) $$\omega(c_2 r) \le \left(1 - \frac{1}{2}\mathfrak{c}\right)\omega(r) + \sigma(r), \qquad 0 < r \le \rho.$$

The same estimate holds if instead of having ess $\sup_{B(y,c_2r)} u_+ \ge 1$, we have ess $\sup_{B(y,c_2r)} u_- \ge 1$ since

$$|\frac{1}{2}\omega(r) - M| = |\text{ess}\sup_{B(y,r)} u| \le N(r).$$

We now apply Lemma 8.23 of [14] which we record verbatim here for the reader's convenience.

LEMMA 63. *Let ω be a nondecreasing function on an interval $(0, R_0]$ satisfying, for all $R \le R_0$, the inequality*

$$\omega(\tau R) \le \gamma\omega(R) + \sigma(R)$$

where σ is also nondecreasing and $0 < \gamma, \tau < 1$. Then, for any $\mu \in (0,1)$ and $R \le R_0$, we have

$$\omega(R) \le C\left(\left(\frac{R}{R_0}\right)^{\alpha}\omega(R_0) + \sigma\left(R^{\mu}R_0^{1-\mu}\right)\right)$$

where $C = C(\gamma)$ and $\alpha = \alpha(\gamma, \tau, \mu)$ are positive constants.

From the lemma we obtain that for any $\mu \in (0, 1)$,

$$\omega(r) \le C\left\{\left(\frac{r}{\rho}\right)^{\alpha}\omega(\rho) + \sigma\left(r^{\mu}\rho^{1-\mu}\right)\right\},$$

where an examination of the proof of Lemma 8.23 in [14] reveals that

$$\begin{aligned}\alpha &= (1-\mu)\frac{\log\gamma}{\log\tau} \\ &= (\mu - 1)\log_{\frac{1}{c_2}}\left(1 - \frac{1}{2}\mathfrak{c}\right) > 0.\end{aligned}$$

Using $\sigma(s) \le \left(\frac{s}{\rho}\right)^{\eta}\sigma(\rho)$ for $0 < s < \rho < 1$, with $s = r^{\mu}\rho^{1-\mu}$, we thus obtain

$$\omega(r) \le C'\left\{\left(\frac{r}{\rho}\right)^{\alpha}\omega(\rho) + \left(\frac{r}{\rho}\right)^{\eta\mu}\sigma(\rho)\right\}, \qquad 0 < r < \rho < 1,$$

for a positive constant C' depending on $\mathfrak{c}$ (and so also on $\|F\|_{\frac{q}{2}}$, $\|\mathbf{G}\|_q$ and $\|\mathbf{H}\|_q$), but not on $\|f\|_{\frac{q}{2}}$, $\|\mathbf{g}\|_q$ or $\|u\|_{L^2}$. Now choose $\mu \in (0, 1)$ so that

$$\alpha = (\mu - 1)\log_{\frac{1}{c_2}}\left(1 - \frac{1}{2}\mathfrak{c}\right) < \eta\mu$$

to obtain

$$(174) \qquad \omega\left(r\right) \leq C(\omega(\rho) + \sigma(\rho)) \left(\frac{r}{\rho}\right)^{\alpha}, \qquad 0 < r \leq \rho < 1,$$

where C depends on $\mathfrak{c}$, $\|F\|_{\frac{q}{2}}$, $\|\mathbf{G}\|_q$ and $\|\mathbf{H}\|_q$. After possible redefinition on a set of measure zero, u is Hölder continuous of order α on $B(y, \rho)$ with respect to the quasimetric d, and we then obtain for any compact set $K \subset \Omega$ that $\|u\|_{C^{\alpha}_{quasi}(K)} < \infty$ where $C^{\alpha}_{quasi}(K)$ is as in (28). If we now invoke the containment condition (11), so that $d\left(x, y\right) \leq C\left|x - y\right|^{\varepsilon}$, we obtain that u is Hölder continuous of order $\varepsilon\alpha$ in the usual Euclidean sense, i.e. $\|u\|_{C^{\varepsilon\alpha}(K)} < \infty$ where $C^{\alpha}(K)$ is as in (8).

This completes the proof of Theorem 8.

REMARK 64. *The Hölder exponent α in (174) satisfies the estimate*

$$\frac{1}{\alpha} \approx \frac{1}{\eta} + \frac{1}{\mathfrak{c}},$$

which is independent *of u, $B\left(y, r\right)$, $\|f\|_{\frac{q}{2}}$ and $\|\mathbf{g}\|_q$.*

In the next theorem, we record the precise constants in estimate (174) using

$$\omega(\rho) + \sigma(\rho) \leq 2N(\rho) + m(\rho) + N(\rho)(\rho^{2\eta}\|F\|_{\frac{q}{2}} + \rho^{\eta}\|G\|_q),$$

and the bound for $N(r)$ given in (148).

THEOREM 65. *Suppose that u is a weak solution to (99) in $B\left(y, \frac{\rho}{c_1}\right)$ with $y \in \Omega$, $0 < \rho < \delta\, dist\,(y, \partial\Omega)$ for sufficiently small δ, where c_1 is as in (148), and that (24), (23), (15) and (17) hold for some $q > Q = \max\{Q^*, 2\sigma'\}$, as well as (20) for some $p > \max\{2\sigma', 4\}$. Then u satisfies the following Hölder estimate:*

$$(175) \qquad \left|u\left(x\right) - u(x')\right| \leq C \left\{ \left(\frac{1}{\left|B\left(y, \frac{\rho}{c_1}\right)\right|} \int_{B\left(y, \frac{\rho}{c_1}\right)} u^2\right)^{\frac{1}{2}} \right.$$

$$\left. + \rho^{2\eta} \|f\|_{L^{\frac{q}{2}}\left(B\left(y, \frac{\rho}{c_1}\right)\right)} + \rho^{\eta} \|\mathbf{g}\|_{L^q\left(B\left(y, \frac{\rho}{c_1}\right)\right)} \right\} \left(\frac{d(x, x')}{\rho}\right)^{\alpha},$$

for $x, x' \in B\left(y, \rho\right)$ where α and C depend on $\mathfrak{c} = C_{Har}^{-1}$, $\|F\|_{\frac{q}{2}}$, $\|\mathbf{G}\|_q$ and $\|\mathbf{H}\|_q$, but are independent of u, $B\left(y, \rho\right)$, $\|f\|_{\frac{q}{2}}$ and $\|\mathbf{g}\|_q$. If in addition the containment condition (11) holds, then we can replace $d\left(x, x'\right)$ with $C\left|x - x'\right|^{\varepsilon}$.

See Remarks 86 and 98 for an explicit estimate of the quasimetric $d\left(x, y\right)$ in the special cases of noninterference and flag balls respectively.

CHAPTER 4

Reduction of the proofs of the rough diagonal extensions of Hörmander's theorem

In the first subsection here, we complete the proof of Theorem 10, our rough analogue of the Fefferman-Phong subellipticity theorem, by establishing the existence of accumulating sequences of Lipschitz cutoff functions as in (20), for the subunit balls $K(y, r)$. Then in the next three subsections, we develop the tools necessary for obtaining Theorem 12, and more generally Theorem 82, our general extension of Hormander's theorem to rough diagonal vector fields. We will collect a variety of conditions, each inherited by locally equivalent families of balls, together in a basic axiom postulating the existence of a prehomogeneous space, with properties appropriate for use in the hypotheses of Theorems 8 and 10. In the proof of Theorem 82, we show that the axiom holds for a particular prehomogeneous space, and thus obtain that it also holds for a locally equivalent symmetric general homogeneous space for which we have established Sobolev and Poincaré inequalities. The three subsections discuss respectively general consequences of the basic axiom, Sobolev and Poincaré inequalities, and finally the proportional subrepresentation inequality, the main condition in the basic axiom. Theorems 12 and 82 then reduce the proofs of Theorems 17 and 20 to demonstrating that for $\delta > 0$ sufficiently small, the flag and noninterference balls form a δ-local prehomogeneous space $\mathcal{B}$ and a δ-local homogeneous space $\mathcal{A}$ respectively, that are adapted to the vector fields, as well as satisfying reverse doubling and size-limiting properties respectively. The strategy is to use these conditions to first obtain the local equivalence of the spaces $\mathcal{B}$ and $\mathcal{A}$ with the subunit balls $\mathcal{K}$, and then to apply Theorem 82. At the end of the section, we turn to proving the sharp technical results, Theorems 23 and 24, again assuming adaptablility of balls to vector fields, but this time establishing directly the "accumulating sequence of Lipschitz cutoff functions" condition (20), as the associated spaces $\mathcal{B}$ and $\mathcal{A}$ need no longer be locally equivalent to the space $\mathcal{K}$ of subunit balls.

1. Accumulating sequences of Lipschitz cutoff functions in annuli

Here we establish the existence of accumulating sequences of Lipschitz cutoff functions as in (20), for the subunit balls $K(y, r)$ provided $\mathcal{Q}(x, \xi)$ is continuous in x and the containment condition (19) holds. First we record the simple facts that if the quadratic form $\mathcal{Q}(x, \xi)$ is continuous, then the subunit metric δ in Definition 4 is an increasing pointwise limit of the Lipschitz continuous metrics δ^{ε} associated to the forms $\mathcal{Q}(x, \xi) + \varepsilon^2 |\xi|^2$, and that the gradients of the metrics δ^{ε} satisfy a pointwise bound in the norm $\|\cdot\|_{\mathcal{Q}}$, uniformly in $\varepsilon > 0$.

LEMMA 66. *Suppose that the quadratic form $\mathcal{Q}(x, \xi)$ is continuous for $x \in \Omega$, and let $\delta(x, y)$ denote the subunit metric in Ω associated to $\mathcal{Q}$ as in Definition 4.*

For $\varepsilon \geq 0$, set $\mathcal{Q}^\varepsilon (x, \xi) = \mathcal{Q}(x, \xi) + \varepsilon^2 |\xi|^2$ and let $\delta^\varepsilon (x, y)$ be the corresponding sub-unit metric. Then for $\varepsilon > 0$, δ^ε is Lipschitz continuous with norm $\frac{1}{\varepsilon}$, and $\delta^\varepsilon (x, y)$ increases to $\delta(x, y)$ as $\varepsilon \to 0$ for all $x, y \in \Omega$, so that δ is lower semicontinuous (and possibly infinite). Moreover, δ^ε satisfies

$$(176) \qquad \|\nabla_x \delta^\varepsilon (x, y)\|_\mathcal{Q} \leq \sqrt{n}, \quad x, y \in \Omega,$$

uniformly in $\varepsilon > 0$, where $\|\cdot\|_\mathcal{Q}$ is as in (14).

REMARK 67. *This inequality was obtained with $\varepsilon = 0$ in a distributional sense in [12] and [11] with a larger constant under the additional hypotheses that $\mathcal{Q}(x, \xi) = \sum_{j=1}^N (X_j (x) \cdot \xi)^2$ is a sum of squares of Lipschitz continuous vector fields X_j, and that δ is bounded above on $\Omega \times \Omega$. More precisely, under these conditions they showed that*

$$(177) \qquad \left| \int f (X_j \varphi) \right| \leq C \int |\varphi (x)|, \quad \varphi \in D(\Omega),$$

where $f(x) = \delta(x, y)$ for y fixed, and $D(\Omega)$ is the space of infinitely differentiable functions with compact support in Ω. This also follows from (176) by first noting that if $f_\varepsilon = \delta^\varepsilon (\cdot, y)$ for fixed $y \in \Omega$ and $\varepsilon > 0$, then f_ε increases monotonically to f, which is now assumed bounded. Thus if $X_j = \sum_{i=1}^n a_{ij}(x) \frac{\partial}{\partial x_i}$, then $|X_j f_\varepsilon| \leq \|\nabla f_\varepsilon\|_\mathcal{Q} \leq \sqrt{n}$ by (176), and so by Lebesgue's dominated convergence theorem,

$$\begin{aligned}
\left| \int f(X_j \varphi) \right| &= \left| \lim_{\varepsilon \to 0} \int f_\varepsilon (X_j \varphi) \right| = \left| \lim_{\varepsilon \to 0} \int (X_j' f_\varepsilon) \varphi \right| \\
&\leq \liminf_{\varepsilon \to 0} \int |(X_j' f_\varepsilon) \varphi| \\
&\leq \liminf_{\varepsilon \to 0} \int |(X_j f_\varepsilon) \varphi| + \sum_{i=1}^n \int \left| \frac{\partial a_{ij}}{\partial x_i} f_\varepsilon \varphi \right| \\
&\leq \liminf_{\varepsilon \to 0} \|X_j f_\varepsilon\|_\infty \int |\varphi| + \sum_{i=1}^n \left\| \frac{\partial a_{ij}}{\partial x_i} \right\|_\infty \|f\|_\infty \int |\varphi| \\
&\leq \left(\sqrt{n} + \sum_{i=1}^n \left\| \frac{\partial a_{ij}}{\partial x_i} \right\|_\infty \|f\|_\infty \right) \int |\varphi|,
\end{aligned}$$

where $\left\| \frac{\partial a_{ij}}{\partial x_i} \right\|_\infty$ is interpreted as in Remark 19. Lemma 66 also yields the following distributional inequality for general quadratic forms $\mathcal{Q}(x, \xi) = \xi' Q(x) \xi$ (different from (177) in the case $\mathcal{Q}$ is a sum of squares of Lipschitz vector fields)

$$(178) \qquad \left| \int f(x) (\nabla' Q \varphi)(x)\, dx \right| \leq \sqrt{n} \int \sqrt{\mathcal{Q}(x, \varphi(x))}\, dx, \quad \varphi \in \mathbf{D}(\Omega),$$

under the additional hypotheses that $\mathcal{Q}(x, \xi)$ is Lipschitz continuous in x and δ is bounded above. Here $\mathbf{D}(\Omega)$ is the space of vector functions $\varphi = (\varphi_1, ..., \varphi_n)$ with $\varphi_i \in D(\Omega)$, $1 \leq i \leq n$. Indeed, fix y and set $f_\varepsilon = \delta^\varepsilon (\cdot, y)$ for $\varepsilon > 0$ as above. Then

$\nabla'Q\varphi$ *is integrable and so by Lebesgue's dominated convergence theorem,*

$$\left|\int f\left(\nabla'Q\varphi\right)\right| = \left|\lim_{\varepsilon\to0}\int f_\varepsilon\left(\nabla'Q\varphi\right)\right| = \left|\lim_{\varepsilon\to0}\int\left(\nabla f_\varepsilon\right)'Q\varphi\right|$$

$$\leq \liminf_{\varepsilon\to0}\int\sqrt{\left(\nabla f_\varepsilon\right)'Q\left(\nabla f_\varepsilon\right)}\sqrt{\varphi'Q\varphi}$$

$$\leq \sqrt{n}\int\sqrt{\varphi'Q\varphi}$$

by (176).

PROOF (OF LEMMA 66). Since δ^ε is a symmetric metric, we have

$$\left|\delta^\varepsilon\left(x,y\right)-\delta^\varepsilon\left(z,y\right)\right|\leq\delta^\varepsilon\left(x,z\right)\leq\frac{|x-z|}{\varepsilon},$$

where the final inequality follows upon considering the following curve joining x to z, that is subunit with respect to $\mathcal{Q}^\varepsilon$:

$$\gamma\left(t\right)=x+\frac{\varepsilon}{|z-x|}t\left(z-x\right),\qquad 0\leq t\leq\frac{|x-z|}{\varepsilon}.$$

Thus δ^ε is Lipschitz continuous with norm $\frac{1}{\varepsilon}$. Next note that $\delta^{\varepsilon_1}\left(x,y\right)\leq\delta^{\varepsilon_2}\left(x,y\right)\leq\delta\left(x,y\right)$ for $0<\varepsilon_2<\varepsilon_1$ follows from $\mathcal{Q}+\varepsilon_1I\succeq\mathcal{Q}+\varepsilon_2I\succeq\mathcal{Q}$. Now suppose that $\delta^\varepsilon\left(x,y\right)<r$ for all $\varepsilon>0$, and choose, for each $0<\varepsilon<1$, a Lipschitz curve $\gamma^\varepsilon\left(t\right)$ satisfying (note that we may stop the curve γ^ε as soon as it hits x)

$$\begin{aligned}\gamma^\varepsilon\left(0\right) &= y,\\\gamma^\varepsilon\left(r\right) &= x,\\\left|\left(\gamma^\varepsilon\right)'\left(t\right)\cdot\xi\right|^2 &\leq \mathcal{Q}\left(\gamma^\varepsilon\left(t\right),\xi\right)+\varepsilon^2\left|\xi\right|^2.\end{aligned}$$

Since $\mathcal{Q}\left(x,\xi\right)$ is bounded in x, the family $\{\gamma^\varepsilon\left(t\right)\cdot\xi\}_{0<\varepsilon<1}$ is equicontinuous in t, and there is a continuous curve $\gamma\left(t\right)$ and a sequence $\{\varepsilon_i\}_{i=1}^\infty$ with $\lim_{i\to\infty}\varepsilon_i=0$, such that $\lim_{i\to\infty}\gamma^{\varepsilon_i}\left(t\right)=\gamma\left(t\right)$ uniformly on $[0,r]$. It now follows that $\gamma\left(t\right)$ is Lipschitz and satisfies

$$\begin{aligned}\gamma\left(0\right) &= y,\\\gamma\left(r\right) &= x,\\\left|\gamma'\left(t\right)\cdot\xi\right|^2 &\leq \mathcal{Q}\left(\gamma\left(t\right),\xi\right),\end{aligned}$$

where the third line follows by considering a fixed difference quotient and letting $i\to\infty$ as follows:

$$\begin{aligned}\left|\frac{\gamma\left(t+h\right)\cdot\xi-\gamma\left(t\right)\cdot\xi}{h}\right|^2 &= \lim_{i\to\infty}\left|\frac{\gamma^{\varepsilon_i}\left(t+h\right)\cdot\xi-\gamma^{\varepsilon_i}\left(t\right)\cdot\xi}{h}\right|^2\\&\leq \liminf_{i\to\infty}\left|\left(\gamma^{\varepsilon_i}\right)'\left(t+c_ih\right)\cdot\xi\right|^2\\&\leq \liminf_{i\to\infty}\left\{\mathcal{Q}\left(\gamma^{\varepsilon_i}\left(t+c_ih\right),\xi\right)+\varepsilon_i^2\left|\xi\right|^2\right\}\\&= \mathcal{Q}\left(\gamma\left(t+ch\right),\xi\right)\end{aligned}$$

upon taking a further subsequence such that $c_i\to c$, and using the uniform convergence of γ^{ε_i} to γ, along with the continuity of $\mathcal{Q}\left(x,\xi\right)$ in x. Now let $h\to0$ and use the continuity of $\mathcal{Q}\left(x,\xi\right)$ in x once more to obtain $\left|\gamma'\left(t\right)\cdot\xi\right|^2\leq\mathcal{Q}\left(\gamma\left(t\right),\xi\right)$. Thus $\delta\left(x,y\right)\leq r$ and this proves that $\delta\left(x,y\right)=\lim_{\varepsilon\to0}\delta^\varepsilon\left(x,y\right)$, and we're done.

We now give an elementary proof of (176). Fix $y \in \Omega$ and set $f_\varepsilon(x) = \delta^\varepsilon(x, y)$. Then f_ε is Lipschitz continuous by what has already been proved. Now fix $x \in \Omega$ and $\varepsilon > 0$ and let $\{\lambda_j^2\}_{j=1}^n$ and $\{\mathbf{v}_j\}_{j=1}^n$ be the eigenvalues and eigenvectors respectively for the matrix $Q^\varepsilon(x) = Q(x) + \varepsilon^2 I$ corresponding to the form $\mathcal{Q}^\varepsilon(x, \xi)$. Then for $1 \leq j \leq n$ and $\beta < 1$, the curve $\gamma_j(t) = x + \beta t \lambda_j \mathbf{v}_j$ is subunit with respect to $\mathcal{Q}^\varepsilon$ for t sufficiently small since

$$
\begin{aligned}
\left(\gamma_j'(t) \cdot \xi\right)^2 &= \left(\beta \lambda_j \mathbf{v}_j \cdot \xi\right)^2 \\
&\leq \beta^2 \sum_{i=1}^n \lambda_i^2 (\mathbf{v}_i \cdot \xi)^2 = \beta^2 \xi' Q^\varepsilon(x) \xi \\
&< \xi' Q^\varepsilon(x + \beta t \lambda_j \mathbf{v}_j) \xi = \xi' Q^\varepsilon(\gamma_j(t)) \xi
\end{aligned}
$$

for small t by the continuity of $Q(x)$. Thus $\delta^\varepsilon(x, x + \beta t \lambda_j \mathbf{v}_j) \leq |t|$ for t small, and since δ^ε is a metric,

$$
\begin{aligned}
|\beta \lambda_j \mathbf{v}_j \cdot \nabla f_\varepsilon(x)| &= \left| \lim_{t \to 0} \frac{f_\varepsilon(x) - f_\varepsilon(x + \beta t \lambda_j \mathbf{v}_j)}{t} \right| \\
&\leq \limsup_{t \to 0} \left| \frac{\delta^\varepsilon(x, x + \beta t \lambda_j \mathbf{e}_j)}{t} \right| \leq 1,
\end{aligned}
$$

for $\beta < 1$ and $1 \leq j \leq n$. Passing to the limit $\beta \to 1$, we obtain

$$
\|\nabla f_\varepsilon(x)\|_{\mathcal{Q}^\varepsilon} = \left(\sum_{i=1}^n |\lambda_i \mathbf{v}_i \cdot \nabla f_\varepsilon(x)|^2 \right)^{\frac{1}{2}} \leq \sqrt{n},
$$

proving (176) since $\|\cdot\|_{\mathcal{Q}} \leq \|\cdot\|_{\mathcal{Q}^\varepsilon}$.

PROPOSITION 68. *The "accumulating sequence of Lipschitz cutoff functions" condition (20) holds in Ω with $p = \infty$ for the subunit balls $K(x, r)$, provided the quadratic form $\mathcal{Q}(x, \xi)$ is continuous for $x \in \Omega$ and the containment condition (19) holds.*

PROOF. To see this, in (20) we take $p = \infty$, $c = \frac{1}{2}$, $N = 2$ and set $\psi_j(x) = \varphi_{3j}(\delta^{\varepsilon_{3j}}(x, y))$ where $\delta^\varepsilon(x, y)$ is the subunit metric associated to the form $\mathcal{Q}^\varepsilon(x, \xi) = \mathcal{Q}(x, \xi) + \varepsilon^2 |\xi|^2$ for $\varepsilon > 0$, and ε_j will be chosen in a moment. Here we define $\varphi_j(t)$ for $j \geq 1$ to vanish for $t \geq r_j$, to equal 1 for $t \leq r_{j+1}$ and to be linear on the interval $[r_{j+1}, r_j]$, where $r_j - r_{j+1} = \frac{r}{5j^2}$, with $r_1 = r$. From the chain rule and (176) we obtain

$$
\|\nabla \psi_j\|_{\mathcal{Q}} \leq \|\nabla \varphi_{3j}\|_\infty \|\nabla_x \delta^{\varepsilon_{3j}}(x, y)\|_{\mathcal{Q}} \leq \frac{5(3j)^2}{r} \sqrt{n}.
$$

Now we've already shown that $\lim_{\varepsilon \to 0} \delta^\varepsilon(x, y) = \delta(x, y)$ on Ω. Since the convergence is monotone and δ is continuous by assumption, $\delta^\varepsilon(\cdot, y) \to \delta(\cdot, y)$ uniformly on compact subsets of Ω (see e.g. Theorem 7.13 in [**37**]).

We now claim that for $0 < s < t < \delta \operatorname{dist}(x, \partial\Omega)$, the annulus $K(x, t) \setminus K(x, s)$ has positive Euclidean thickness, in fact with C and ε as in (11),

$$
\operatorname{dist}(K(x, s), K(x, t)^c) \equiv \inf_{w \in K(x, s), z \notin K(x, t)} |w - z| \geq \frac{(t - s)^{\frac{1}{\varepsilon}}}{C} > 0.
$$

To see this, take $w \in K(x, s)$ and $z \notin K(x, t)$ so that $\delta(x, w) < s$ and $\delta(x, z) \geq t$. Since δ is a metric, we have $\delta(w, z) > t - s$. But then by (19),

$$D\left(w, \frac{(t-s)^{\frac{1}{\varepsilon}}}{C}\right) \subset K(w, t - s),$$

and so $z \notin D\left(w, \frac{(t-s)^{\frac{1}{\varepsilon}}}{C}\right)$ yields $|w - z| \geq \frac{(t-s)^{\frac{1}{\varepsilon}}}{C}$ as required.

So by taking $\varepsilon_j > 0$ small enough, the Lipschitz function $x \to \varphi_j\left(\delta^{\varepsilon_j}(x, y)\right)$ will have its derivative supported in the annulus $K(y, r_{j-1}) \backslash K(y, r_{j+2})$ for $j \geq 1$. Since these annuli are pairwise disjoint and contained in $K(y, r) \backslash K\left(y, \frac{r}{2}\right)$ for $j \in 3\mathbb{N}$ (since $\sum_{j=1}^{\infty} \frac{1}{5j^2} < \frac{1}{2}$), the sequence of functions $\{\psi_j\}_{j=1}^{\infty}$, with $\psi_j(x) = \varphi_{3j}\left(\delta^{\varepsilon_{3j}}(x, y)\right)$, is now easily seen to satisfy (20) with $p = \infty$, $c = \frac{1}{2}$ and $N = 2$. This completes the proof of the proposition.

2. The axiom

Let $\mathcal{X}$ be a collection of *continuous* vector fields $X_j = a_j(x)\frac{\partial}{\partial x_j}$, $1 \leq j \leq n$, on Ω with $a_1(x) \equiv 1$ and set $A_j(x, r) = \int_0^r a_j(x_1 + t, x_2, ..., x_n)\,dt$, $1 \leq j \leq n$, provided the segment joining x and $x + (r, 0, ..., 0)$ lies in Ω. In this subsection we introduce our basic axiom regarding $\mathcal{X}$, and derive some general consequences from it.

AXIOM 1. *There is a prehomogeneous space on Ω of open subsets $\mathcal{B} = \{B(x, r)\}_{x \in \Omega, 0 < r < \infty}$ as in Definition 34 (i.e., $(56), (58), (60)$ and (62) hold), satisfying the convex hull equivalence (12),*

$$(179) \qquad coB(x, r) \subset B(x, Cr), \qquad x \in \Omega,\ 0 < r < \delta\ dist(x, \partial\Omega),$$

and the doubling condition

$$(180) \qquad \left|\widetilde{B}(x, 2r)\right| \leq C\left|\widetilde{B}(x, r)\right|, \qquad x \in \Omega,\ 0 < r < \delta\ dist(x, \partial\Omega),$$

where $\widetilde{B}(x, r) = \prod_{j=1}^{n}[x_j - B_j(x, r), x_j + B_j(x, r)]$ is the smallest closed rectangle containing $B(x, r)$ as defined in (89). Moreover, the prehomogeneous space of open sets $\mathcal{B}$ is related to the collection of vector fields $\mathcal{X}$ by the following two properties:

  (1) *The side-limiting inequality (note that $A_1(x, r) \equiv r$):*

$$(181) \qquad C^{-1} \leq \frac{B_1(x, r)}{r} \leq C, \qquad x \in \Omega,\ 0 < r < \delta\,dist(x, \partial\Omega).$$

  (2) *The proportional subrepresentation inequality: for every $0 < \beta < 1$, there is a positive constant C_β such that for all $y \in \Omega$ and $0 < r < \delta\,dist(y, \partial\Omega)$,*

$$(182) \qquad |f(x)| \leq C_\beta \int_{B(y, Cr)} |\nabla_a f(z)|\,\mathcal{K}(x, z)\,dz,$$

  whenever f is Lipschitz continuous, $x \in B(y, r)$ and $|\mathcal{Z}(f) \cap B(y, r)| \geq \beta|B(y, r)|$ where $\mathcal{Z}(f)$ denotes the zero set of f. Here, $\mathcal{K}(x, z)$ is given by

$$(183) \qquad \mathcal{K}(x, z) = \frac{d(x, z)}{|B(x, d(x, z))|},$$

where

$$(184) \qquad d(x,y) = \inf \{ r > 0 : y \in B(x,r) \}.$$

Note that (184) defines the quasimetric associated to the prehomogeneous space as in Remark 35.

REMARK 69. *If* $\mathcal{B} = \{ B(x,r) \}_{x \in \Omega, 0 < r < \infty}$ *is a prehomogeneous space of open subsets on* Ω, *then the doubling condition (180) on the rectangles* $\widetilde{B}(x,r)$ *implies doubling for their half sidelengths* $B_j(x,r)$:

$$B_j(x,2r) \leq CB_j(x,r), \qquad x \in \Omega, \, 0 < r < \delta \, dist(x, \partial\Omega).$$

for $1 \leq j \leq n$. *Indeed, using only weak monotonicity (60) and doubling (180), we have with* $0 < c < 1$ *the constant in (60),*

$$B_j\left(x, c^{-1}r\right) = \frac{2^{-n} \left| \widetilde{B}\left(x, c^{-1}r\right) \right|}{\prod_{i \neq j} B_i\left(x, c^{-1}r\right)} \leq \frac{C' 2^{-n} \left| \widetilde{B}(x,r) \right|}{\prod_{i \neq j} B_i(x,r)} = C' B_j(x,r).$$

Let us show that the properties (56), (58), (60), (62), (179), (180), (181) and (182) of Axiom 1 persist for Ω-locally equivalent families of sets $\{ B^*(x,r) \}_{x \in \Omega, 0 < r < \infty}$, possibly with a smaller positive constant δ_0. Recall from subsection 2.3 that the families of sets $\{ B(x,r) \}_{x \in \Omega, 0 < r < \infty}$ and $\{ B^*(x,r) \}_{x \in \Omega, 0 < r < \infty}$ are Ω-*locally equivalent* if there are positive constants C and δ such that
(185)
$$B(x,r) \subset B^*(x,Cr) \text{ and } B^*(x,r) \subset B(x,Cr), \qquad x \in \Omega, \, 0 < Cr < \delta \, dist(x, \partial\Omega).$$

It is this flexibility of the properties in Axiom 1 that makes the axiom useful in dealing with the prehomogeneous space of flag balls in section 5. Now it is obvious that (56), (60), (58), (62), (179) and (180) are inherited by an Ω-locally equivalent family of sets. From Remark 69, the side limiting inequality (181) is also inherited. Next we turn to the proportional subrepresentation inequality. We note that in the presence of the engulfing and doubling properties, (56) and (62), $d(x,y)$ in (184) may be replaced by any equivalent function $\widetilde{d}(x,y)$, since then

$$\frac{d(x,z)}{|B(x,d(x,z))|} \approx \frac{\widetilde{d}(x,z)}{\left| B\left(x, \widetilde{d}(x,z)\right) \right|}.$$

LEMMA 70. *Let* $\mathcal{B} = \{ B(x,r) \}_{x \in \Omega, 0 < r < \infty}$ *and* $\mathcal{B}^* = \{ B^*(x,r) \}_{x \in \Omega, 0 < r < \infty}$ *be families of sets satisfying the engulfing and doubling properties, (56) and (62), and suppose that they are* Ω-*locally equivalent in the sense that (185) holds. If the proportional subrepresentation inequality (182) holds for* $\mathcal{B} = \{ B(x,r) \}_{x \in \Omega, 0 < r < \infty}$, *then it holds for* $\mathcal{B}^*$.

PROOF. If (185) holds, then the corresponding kernels $\mathcal{K}$ and $\mathcal{K}^*$ are equivalent. Suppose we are given $B^*(y,r)$ with $0 < Cr < \delta \, dist(x, \partial\Omega)$. Let $x \in B^*(y,r)$ and let f Lipschitz satisfy $|\mathcal{Z}(f) \cap B^*(y,r)| \geq \beta^* |B^*(y,r)|$, $0 < \beta^* < 1$. By the equivalence of sets (185), $B(y,Cr) \supset B^*(y,r)$. Then we have $x \in B(y,Cr)$ and $|\mathcal{Z}(f) \cap B(y,Cr)| \geq \beta |B(y,Cr)|$ for some $0 < \beta < 1$. By (182) for the sets B, and (185) again, we obtain

$$|f(x)| \leq C_\beta \int_{B(y,CCr)} |\nabla_a f(z)| \, \mathcal{K}(x,z) \, dz$$

$$\leq CC_\beta \int_{B^*(y,CC^2r)} |\nabla_a f(z)| \, \mathcal{K}^*(x,z) \, dz,$$

which is (182) for the sets B^*.

Remark 35 and Lemma 37, together with the above discussion, now immediately yield the following lemma which will prove useful in connection with the flag balls, as they enjoy only the weak monotonicity property (60).

LEMMA 71. *Suppose that* $\{B(x,r)\}_{x \in \Omega, 0 < r < \infty}$ *is a prehomogeneous space on* Ω *satisfying Axiom 1. Let* $\{B^*(x,r)\}_{x \in \Omega, 0 < r < \infty}$ *be the general homogeneous space with symmetric quasimetric* $d^*_{sym}(x,y)$ *as in Lemma 37, arising from the quasimetric* $d(x,y)$ *in Remark 35. Then the properties (56), (58), (60), (62), (179), (180), (181) and (182) of Axiom 1 are satisfied with the balls* $B^*(x,r)$ *in place of the sets* $B(x,r)$. *Moreover, for some positive constant* C, *we have*

$$B\left(x, C^{-1}r\right) \subset B^*(x,r) \subset B(x,Cr), \quad 0 < r < \infty.$$

In preparation for proving Sobolev and Poincaré inequalities in the next subsection, we show that for a general homogeneous space, the proportional subrepresentation inequality (182) implies the standard subrepresentation inequality for Lipschitz continuous f: there are positive constants δ, C, C_0 such that for all $y \in \Omega$ and $0 < r < \delta \, dist(y, \partial\Omega)$,

$$(186) \qquad |f(x) - f_{B(y,r)}| \leq C \int_{B(y,C_0 r)} |\nabla_a f(z)| \, \mathcal{K}(x,z) \, dz$$

$$+ \frac{Cr}{|B(y,C_0 r)|} \int_{B(y,C_0 r)} |\nabla_a f(z)| \, dz,$$

for $x \in B(y,r)$, where $f_{B(y,r)} = \frac{1}{|B(y,r)|} \int_{B(y,r)} f$ is the average value of f on $B(y,r)$, and the fractional integral kernel $\mathcal{K}$ is given by (183). The sharper version of (186) with $C_0 = 1$ is often available in special cases; see for example Remark 81 in section 4.4.1. Moreover, it is sometimes possible to drop the second term on the right side of (186); see for example the discussion about reverse doubling of order 1 in Remark 45 of section 2.3.

LEMMA 72. *If the proportional subrepresentation inequality (182) holds with* $\beta = \frac{1}{2}$ *for a general homogeneous space* $\mathcal{B} = \{B(x,r)\}_{x \in \Omega, 0 < r < \infty}$, *then the standard subrepresentation inequality (186) holds.*

PROOF. Fix $y \in \Omega$ and $0 < r < \delta \, dist(y, \partial\Omega)$, and let ω be the median value of f on $B(y,r)$. Set

$$f_+(x) = \begin{cases} f(x) - \omega, & \text{for} \quad f(x) > \omega \\ 0 & \text{for} \quad f(x) \leq \omega \end{cases},$$

$$f_-(x) = \begin{cases} \omega - f(x), & \text{for} \quad f(x) < \omega \\ 0 & \text{for} \quad f(x) \geq \omega \end{cases},$$

for $x \in B(y, C_0 r)$, so that $f(x) - \omega = f_+(x) - f_-(x)$. Then

$$
\begin{aligned}
\left| f(x) - f_{B(y,r)} \right| &\leq \left| f(x) - \omega \right| + \left| f_{B(y,r)} - \omega \right| \\
&\leq \left| f_+(x) \right| + \left| f_-(x) \right| \\
&\quad + \frac{1}{|B(y,r)|} \int_{B(y,r)} \left| f_+(w) \right| \\
&\quad + \frac{1}{|B(y,r)|} \int_{B(y,r)} \left| f_-(w) \right|.
\end{aligned}
$$

Since $\left| \mathcal{Z}(f_\pm) \cap B(y,r) \right| \geq \frac{1}{2} |B(y,r)|$ we can apply (182) to each of f_+ and f_- with $\beta = \frac{1}{2}$, and use $|\nabla_a f_\pm| \leq |\nabla_a f|$ almost everywhere, to obtain

(187)

$$
\begin{aligned}
\left| f(x) - f_{B(y,r)} \right| &\leq 2 C_{\frac{1}{2}} \int_{B(y,C_0 r)} |\nabla_a f(z)| \, \mathcal{K}(x,z) \, dz \\
&\quad + 2 C_{\frac{1}{2}} \int_{B(y,C_0 r)} |\nabla_a f(z)| \left\{ \frac{1}{|B(y,r)|} \int_{B(y,r)} \mathcal{K}(w,z) \, dw \right\} dz \\
&\leq 2 C_{\frac{1}{2}} \int_{B(y,C_0 r)} |\nabla_a f(z)| \, \mathcal{K}(x,z) \, dz \\
&\quad + 2 C C_{\frac{1}{2}} \frac{r}{|B(y,C_0 r)|} \int_{B(y,C_0 r)} |\nabla_a f(z)| \, dz,
\end{aligned}
$$

since (97) implies

$$
\frac{1}{|B(y,r)|} \int_{B(y,r)} \mathcal{K}(w,z) \, dw \leq C \frac{r}{|B(y,r)|}, \qquad x \in B(y,r), z \in B(y, C_0 r).
$$

This completes the proof of the lemma.

For use in the next subsection, we also record the following two standard properties of homogeneous spaces.

LEMMA 73. *Suppose $\{B(x,r)\}_{x \in \Omega, 0 < r < \infty}$ is the family of balls in a homogeneous space on Ω.*

(a) *There are positive constants C and D such that*

$$
|B(x,r)| \leq C \left(\frac{r}{t} \right)^D |B(y,t)|, \qquad \text{whenever } B(x,r) \supset B(y,t).
$$

(b) *If $B(x,r) \cap B(y,t) \neq \phi$ and $r \approx t$, then $|B(x,r)| \approx |B(y,t)|$.*

3. The Sobolev and Poincaré inequalities

The following Poincaré inequality clearly implies both the Sobolev inequality (15) and the Poincaré inequality (17) that are needed to invoke Theorem 8 or Theorem 10.

PROPOSITION 74. *Let $\{B(x,r)\}_{x \in \Omega, 0 < r < \infty}$ be the family of quasimetric balls in a symmetric general homogeneous space on Ω with quasimetric d, and suppose that the standard subrepresentation inequality (186) holds. Then we have the Poincaré*

*inequality: there are positive constants C, C_0 such that for sufficiently small $\delta > 0$
and all $y \in \Omega$ and $0 < r < \delta \, dist\,(y, \partial\Omega)$,*

$$\left(\frac{1}{|B(y,r)|} \int_{B(y,r)} |f - f_{B(y,r)}|^q \right)^{\frac{1}{q}} \leq Cr \left(\frac{1}{|B(y,C_0 r)|} \int_{B(y,C_0 r)} |\nabla_a f|^p \right)^{\frac{1}{p}},$$

*for all f Lipschitz on $B(y, C_0 r)$, and $1 < p < q < \infty$ where $\frac{1}{q} = \frac{1}{p} - \frac{1}{D}$ and D is
the doubling exponent in Lemma 73 (a).*

This result remains true for $p = 1$; see Remark 77 at the end of the proof.

PROOF. There are several possible ways to proceed, and we have chosen to
deduce the Poincaré estimate from the results in [**41**] about integral operators of
potential type in a symmetric general homogeneous space. We recall the context.
Let $\rho(\cdot, \cdot)$ be a symmetric Lebesgue measurable quasimetric on an open subset Ω
of $\mathbb{R}^n$, and let $B(x, r)$ denote the ρ-ball with center x and radius r. Suppose that
Lebesgue measure satisfies the doubling condition

$$(188) \qquad |B(x, 2r)| \leq C \, |B(x, r)|, \qquad 0 < r < \infty,$$

with C independent of x and r. Thus the triple (Ω, ρ, dx) (where dx denotes
Lebesgue measure) is a general homogeneous space in the sense of Definition 33.
Following [**41**], we consider integral operators $T : g \to Tg$ which have the form

$$(189) \qquad Tg(x) = \int_{B_0} g(z) \, \mathcal{K}(x, z) \, dz,$$

where B_0 is a fixed ball (with respect to the given quasimetric ρ) and $\mathcal{K}(x, z)$ is
a nonnegative kernel. Given a kernel $\mathcal{K}$ and any ρ-quasimetric ball B with radius
$r(B)$, let

$$(190) \qquad \phi(B) = \sup\{\mathcal{K}(x, z) : x, z \in B, \rho(x, z) \geq c_1 r(B)\},$$

where c_1 is a sufficiently small positive constant which depends only on the quasimetric constant of ρ (i.e., only on the analogue for ρ of the constant κ in (54)). In
fact, if we denote the quasimetric constant of ρ by κ, it is shown on p. 820 of [**41**]
that the choice

$$(191) \qquad c_1 = \frac{1}{9\kappa^4}$$

is possible. In this context, we will use the following special case of Theorem 3B
of [**41**] (by choosing the weights w and v there to have the constant values $1/|B_0|$
and $r(B_0)^p / |B_0|$, respectively).

LEMMA 75. *Suppose that $1 < p < q < \infty$, T is given by (189) for a fixed ρ-ball
B_0, ϕ is defined by (190), and κ is the quasimetric constant of ρ. Let B_0^* denote
the ball with the same center as B_0 and radius $13\kappa^5 r(B_0)$. If there is a constant C
so that*

$$(192) \qquad \phi(B) \, |B| \left(\frac{|B_0|}{|B|} \right)^{\frac{1}{p} - \frac{1}{q}} \leq Cr(B_0)$$

for all ρ-balls $B \subset B_0^*$, then

$$\left(\frac{1}{|B_0|} \int_{B_0} |Tg(x)|^q \, dx \right)^{1/q} \leq C_N r(B_0) \left(\frac{1}{|B_0|} \int_{B_0} |g(x)|^p \, dx \right)^{1/p}$$

with c_N depending only on p, q, κ and the constants in (188) and (192).

REMARK 76. *The results of [41] are derived with the extra assumption that annuli $B(x, R) \setminus B(x, r), 0 < r < R$, are not empty. In fact, this assumption is not needed for Theorem 3B of [41], as can be seen by going through its proof and observing that since empty annuli contribute nothing to T, the only balls B which actually arise in the proof are ones for which $\phi(B)$ is well-defined, i.e., ones which give rise to nonempty annuli.*

In order to apply Lemma 75 to (186), we fix a ball $B_0 = B_0(x_0, r_0)$ with $x_0 \in \Omega$ and

$$0 < 13\kappa^5 r < r_0 < \delta \, dist(x_0, \partial\Omega),$$

and rewrite (186) in the form

$$(193) \qquad \left| f(x) - f_{B(y,r)} \right| \leq CT \left(\chi_{B(y,Cr)} \left| \nabla_a f(z) \right| \right)(x) + C\mathcal{A}, \qquad x \in B(y, r),$$

for $B(y, C_0 r) \subset B_0$, where $\mathcal{A} = \frac{r}{|B(y,C_0 r)|} \int_{B(y,C_0 r)} |\nabla_a f(x)| \, dx$ and the operator T is given by (189) with

$$\mathcal{K}(x, z) = \frac{d(x, z)}{|B(x, d(x, z))|}.$$

(The doubling property of Lebesgue measure for quasimetric balls implies that $\mathcal{K}(x, z) \approx \mathcal{K}(z, x)$, although we shall not use this fact.) We must now verify (192). To estimate the functional $\phi(B)$ in (190), fix a ball B of radius $r(B)$ with $B \subset B_0^*$, for B_0^* as in Lemma 75, and let x and z lie in B with $d(x, z) \geq c_1 r(B)$ with c_1 as in (190). Then

$$\mathcal{K}(x, z) = \frac{d(x, z)}{|B(x, d(x, z))|} \leq \frac{2\kappa r(B)}{|B(x, c_1 r(B))|} \leq C \frac{r(B)}{|B|}$$

by Lemma 73 (b) applied to the balls B and $B(x, c_1 r(B))$. Taking the supremum over such x and z yields

$$\phi(B) \leq C \frac{r(B)}{|B|}.$$

We also have by Lemma 73 (a) that

$$\left(\frac{|B_0|}{|B|} \right)^{\frac{1}{p} - \frac{1}{q}} \leq \left(\frac{|B_0^*|}{|B|} \right)^{\frac{1}{p} - \frac{1}{q}} \leq C \frac{r(B_0)}{r(B)}$$

since $p^{-1} - q^{-1} = D^{-1}$ and $B \subset B_0^*$ by hypothesis. Taking the product of these estimates leads to (192) in the form

$$\phi(B) |B| \left(\frac{|B_0|}{|B|} \right)^{\frac{1}{p} - \frac{1}{q}} \leq C^2 r(B_0)$$

for all $B \subset B_0^*$.

Let $B(y, r)$ be a ball with $y \in \Omega$ and $0 < r < \delta \, dist(y, \partial\Omega)$. Then if $p > 1$ and $p^{-1} - q^{-1} = D^{-1}$, we obtain by combining (193) and Lemma 75 with $B_0 =$

$B\left(y,C_{0}r\right)$, that

$$\left(\frac{1}{\left|B\left(y,r\right)\right|}\int_{B\left(y,r\right)}\left|f\left(x\right)-f_{B\left(y,r\right)}\right|^{q}\,dx\right)^{\frac{1}{q}}$$

$$\leq\ C\left(\frac{\left|B_{0}\right|}{\left|B\left(y,r\right)\right|}\right)^{\frac{1}{q}}\left(\frac{1}{\left|B_{0}\right|}\int_{B_{0}}T\left(\left|\nabla_{a}f\right|\chi_{B_{0}}\right)\left(x\right)^{q}\,dx\right)^{\frac{1}{q}}+C\mathcal{A}$$

$$\leq\ C\left(\frac{\left|B_{0}\right|}{\left|B\left(y,r\right)\right|}\right)^{\frac{1}{q}}C_{N}\,r\left(B_{0}\right)\left(\frac{1}{\left|B_{0}\right|}\int_{B_{0}}\left|\nabla_{a}f\left(x\right)\right|^{p}\,dx\right)^{\frac{1}{p}}+C\mathcal{A}$$

$$\leq\ Cr\left(\frac{1}{\left|B\left(y,Cr\right)\right|}\int_{B\left(y,Cr\right)}\left|\nabla_{a}f\left(x\right)\right|^{p}\,dx\right)^{\frac{1}{p}},$$

since $\left|B_{0}\right|=\left|B\left(y,C_{0}r\right)\right|\leq C\left|B\left(y,r\right)\right|$ and $\mathcal{A}\leq r\left(\frac{1}{\left|B\left(y,C_{0}r\right)\right|}\int_{B\left(y,C_{0}r\right)}\left|\nabla_{a}f\left(x\right)\right|^{p}\,dx\right)^{\frac{1}{p}}$ by Hölder's inequality. This completes the proof of Proposition 74.

REMARK 77. *Proposition 74 remains valid for $p=1$. The argument in case $p=1$ is based on a weak-type analogue of Lemma 75, i.e., it is based on an estimate like the one in Lemma 75 but with the Lorentz norm $\left\|Tg/\left|B_{0}\right|\right\|_{L^{q,\infty}\left(B_{0}\right)}$ on the left, and it also uses a truncation argument involving the differential operator ∇_{a}. For details, we refer for example to Theorem 2 of [10], from which the desired estimate follows as a special case.*

4. Adapted vector fields and prehomogeneous spaces

The purpose of this subsection is to formulate and prove a general subellipticity theorem for a collection $\mathcal{X}$ of diagonal Lipschitz continuous vector fields, Theorem 82 below, that obtains subellipticity for $\mathcal{X}$ when $\mathcal{X}$ is adapted to a prehomogeneous space in a natural way - see Definition 78 below. The key to obtaining the proportional subrepresentation inequality (182) of Axiom 1 from Definition 78 is the lemma below that controls the difference of two proportional averages of a function f by an appropriately normalized average of $\nabla_{a}f$. In order to state the lemma, we recall some definitions.

Let $a\left(x\right)=\left(1,a_{2}\left(x\right),\ldots,a_{n}\left(x\right)\right)$, $x=\left(x_{1},\ldots,x_{n}\right)$, where the a_{i} are nonnegative on Ω. We assume that the a_{i} satisfy the doubling condition in x_{1} uniformly in $x_{2},\ldots,x_{n}$, i.e. that there is a constant C_{d} so that if I and J are one-dimensional intervals with $J\subset 5I$, then

$$\int_{J}a_{i}\left(x_{1},x_{2},\ldots,x_{n}\right)dx_{1}\leq C_{d}\int_{I}a_{i}\left(x_{1},x_{2},\ldots,x_{n}\right)dx_{1}$$

for each i uniformly in $x_{2},\ldots,x_{n}$. We refer to the constant C_{d} as the doubling constant of a. The blow-up factor 5 is present for technical convenience. We also assume that $a_{i}\left(x_{1},x_{2},\ldots,x_{n}\right)$ is Lipschitz continuous in $x_{2},\ldots,x_{n}$ uniformly in x_{1} with Lipschitz constant C_{L}:

$$\left|a_{i}\left(x_{1},x_{2},\ldots,x_{n}\right)-a_{i}\left(x_{1},x_{2}',\ldots,x_{n}'\right)\right|\leq C_{L}\max_{j=2,\ldots,n}\left|x_{j}-x_{j}'\right|.$$

Given $i=2,\ldots,n$, $x\in\mathbb{R}^{n}$ and $t\in\left(-\infty,\infty\right)$, $A_{i}\left(x,t\right)$ is defined by

$$A_{i}\left(x,t\right)=\int_{0}^{t}a_{i}\left(x_{1}+s,x_{2},\ldots,x_{n}\right)ds,$$

provided the segment joining x and $x + (t, 0, ..., 0)$ lies in Ω. Note that $A_i(x, t)$ has the same sign as t.

The values $A_i(x, r)$ are related to integral curves of the vector fields in the linear span of $\left\{ \frac{\partial}{\partial x_1}, a_2 \frac{\partial}{\partial x_2}, \ldots, a_n \frac{\partial}{\partial x_n} \right\}$. In order to guarantee the existence and uniqueness of these integral curves, we assume that the a_i are continuous in x as well as Lipschitz continuous in $x_2, ..., x_n$ (see e.g. Exercise 26 on page 171 and Exercise 28 on page 119 of [**37**]). In fact, for $x \in \Omega$, $u = (u_2, \ldots, u_n) \in R^{n-1}$ and t in an open interval containing 0, let $\gamma_u(x, t)$ be the unique integral curve of the vector field $(1, u_2 a_2, \ldots, u_n a_n)$ with $\gamma_u(x, 0) = x$. Thus $\gamma_u(x, t)$ satisfies

$$\gamma_u'(x, t) = (1, u_2 a_2(\gamma_u(x, t)), \ldots, u_n a_n(\gamma_u(x, t))) \text{ and } \gamma_u(x, 0) = x,$$

and so,

$$\gamma_u(x, t) = (x_1 + t, \gamma_{u2}(x, t), \ldots, \gamma_{un}(x, t)) \quad \text{where}$$

$$\gamma_{ui}(x, t) = x_i + u_i \int_0^t a_i(\gamma_u(x, s)) \, ds, \quad i = 2, \ldots, n.$$

When $u = 0$, $\gamma_0(x, t)$ is simply the segment $(x_1 + t, x_2, \ldots, x_n)$, and consequently

$$A_i(x, t) = \int_0^t a_i(\gamma_0(x, s)) \, ds, \quad i = 2, \ldots, n.$$

In the next lemma, we will vary the initial points x of the curves $\gamma_u(x, t)$ for special fixed values of u to help estimate integral averages of functions. This technique, but instead with x fixed and u varying, was first introduced by Franchi in [**8**].

Let $\mathcal{B} = \{B(x, r) : x \in \Omega, 0 < r < \infty\}$ be a prehomogeneous space on Ω and let $\widetilde{\mathcal{B}}$ be the collection of smallest closed rectangles

$$\widetilde{B}(x, r) = \prod_{k=1}^{n} [x_k - B_k(x, r), x_k + B_k(x, r)]$$

containing $B(x, r)$, as defined in (89). Let the a-gradient $\nabla_a f$ of a function f be denoted as usual by

$$\nabla_a f(x) = \left(\frac{\partial}{\partial x_1} f(x), a_2(x) \frac{\partial}{\partial x_2} f(x), \ldots, a_n(x) \frac{\partial}{\partial x_n} f(x) \right).$$

Given $\alpha > 0$ and $\mathcal{E} \subset \widetilde{B}(x, r)$, we say $\mathcal{E}$ is an α-*proportional subset* of $\widetilde{B}(x, r)$ if $|\mathcal{E}| \geq \alpha \left| \widetilde{B}(x, r) \right|$. The key assumption in our lemma is the following definition.

DEFINITION 78. *A collection of vector fields* $\mathcal{X} = \{X_j(x)\}_{j=1}^n$ *on* Ω, *where* $X_1 = \frac{\partial}{\partial x_1}$, $X_2(x) = a_2(x) \frac{\partial}{\partial x_2}$, ... , $X_n(x) = a_n(x) \frac{\partial}{\partial x_n}$, *is said to be* adapted *on* Ω *to a prehomogeneous space* $\mathcal{B}$ *on* Ω *(or vice versa,* $\mathcal{B}$ *is* adapted *on* Ω *to* $\mathcal{X}$*) if there is* $\delta > 0$ *such that*

(1) *For all* $x \in \Omega$ *with* $0 < r < \delta \, dist(x, \partial\Omega)$, *we have*

$$C^{-1} \leq \frac{B_1(x, r)}{r} \leq C,$$

(2) *For every* $0 < \alpha < 1$ *there is a positive constant* $\varepsilon = \varepsilon(\alpha)$ *such that for all balls* $B(x, r) \in \mathcal{B}$ *with* $x \in \Omega$ *and* $0 < r < \delta \, dist(x, \partial\Omega)$, *there is an* α-*proportional subset* $\mathcal{E}$ *of* $\widetilde{B}(x, r)$ *satisfying*

$$A_j(z, r) \geq \varepsilon B_j(x, r), \quad for \ z \in \mathcal{E}, 1 \leq j \leq n.$$

The next lemma is basic.

LEMMA 79. *Assume that a_i is nonnegative and continuous for x in Ω, doubling in x_1 uniformly in $x_2, \ldots, x_n$ and Lipschitz continuous in $x_2, \ldots, x_n$ uniformly in x_1. Let $X_1 = \frac{\partial}{\partial x_1}$, $X_2(x) = a_2(x) \frac{\partial}{\partial x_2}$, $\ldots$, $X_n(x) = a_n(x) \frac{\partial}{\partial x_n}$ and suppose that the collection of vector fields $\mathcal{X} = \{X_j(x)\}_{j=1}^n$ is adapted on Ω to a prehomogeneous space $\mathcal{B}$ on Ω. Then for any Lipschitz continuous function f on $\widetilde{B}(y, r)$ with $y \in \Omega$ and $0 < r < \delta \, dist(y, \partial\Omega)$, and any α-proportional subsets G and H of $\widetilde{B}(y, r)$ with $0 < \alpha < 1$, we have*

$$\left| \frac{1}{|G|} \int_G f(z)\, dz - \frac{1}{|H|} \int_H f(z)\, dz \right| \leq C \frac{r}{\left| \widetilde{B}(y, r) \right|} \int_{\widetilde{B}(y, r)} |\nabla_a f(z)|\, dz,$$

where the constant C depends on C_d, C_L, δ, α and the constants C and $\varepsilon(\alpha)$ in Definition 78.

PROOF. We assume that $y = 0$ and $0 < r < \delta \, dist(0, \partial\Omega)$ are fixed, and let $B_j(r) = B_j(0, r)$. We rearrange variables so that

(194)
$$B_j(r) \leq B_{j-1}(r) \qquad \text{for } 2 \leq j \leq n.$$

Fix α with $0 < \alpha < 1$. By condition 2 of Definition 78, there exists $\varepsilon > 0$ depending only on α, and a set $\mathcal{E} \subset \widetilde{B}(y, r)$ with $|\mathcal{E}| \geq \alpha \left| \widetilde{B}(y, r) \right|$ such that

(195) $A_j(x, r) = \int_0^r a_j(x_1 + t, x_2, \ldots, x_n)\, dt \geq \varepsilon B_j(r)$ if $(x_1, x_2, \ldots, x_n) \in \mathcal{E}$

for $j = 1, 2, \ldots, n$. By comparing each of the integral averages of f over G and over H to the average of f over $\mathcal{E}$, we may suppose that $H = \mathcal{E}$. Thus the estimate in (195) holds if $(x_1, x_2, \ldots, x_n) \in H$.

Given points $P \in \widetilde{B}(0, r)$ and $Q \in H = \mathcal{E}$, we denote

$$P = (p_1, p_2, \ldots, p_n), \quad Q = (q_1, q_2, \ldots, q_n),$$

and define points $\{P_k\}_{k=1}^{n+1}$ by $P_1 = P$, $P_{n+1} = Q$ and

$$P_k = (q_1, q_2, \ldots, q_{k-1}, p_k, \ldots, p_n), \quad k = 2, \ldots, n.$$

Then

$$
\begin{aligned}
|f(P) - f(Q)| &\leq \sum_{k=1}^n |f(P_k) - f(P_{k+1})| \\
&= |f(p_1, p_2, \ldots, p_n) - f(q_1, p_2, \ldots, p_n)| \\
&\quad + \sum_{k=2}^n |f(q_1, q_2, \ldots, q_{k-1}, p_k, \ldots, p_n) \\
&\qquad\qquad - f(q_1, q_2, \ldots, q_k, p_{k+1}, \ldots, p_n)| \, .
\end{aligned}
$$

We will estimate each term $|f(P_k) - f(P_{k+1})|$ separately. In the simple case $k = 1$, we have

$$|f(P_1) - f(P_2)| \leq \int_{-B_1(r)}^{B_1(r)} \left| \frac{\partial f}{\partial x_1}(t, p_2, \ldots, p_n) \right| dt,$$

since $p_1, q_1 \in [-B_1(r), B_1(r)]$.

Now let $2 \leq k \leq n$ and consider the two integral curves emanating from P_k and P_{k+1} given by

$$\gamma_{0\ldots 0 u_k 0 \ldots 0}\left(P_k, t\right) =$$

$$\left(q_1 + t, q_2, \ldots, q_{k-1}, p_k + u_k \int_0^t a_k\left(\gamma_{0\ldots 0 u_k 0 \ldots 0}\left(P_k, s\right)\right) ds, p_{k+1}, \ldots, p_n \right),$$

and

$$\gamma_{0\ldots 0 u_k 0 \ldots 0}\left(P_{k+1}, t\right) =$$

$$\left(q_1 + t, q_2, \ldots, q_{k-1}, q_k + u_k \int_0^t a_k\left(\gamma_{0\ldots 0 u_k 0 \ldots 0}\left(P_{k+1}, s\right)\right) ds, p_{k+1}, \ldots, p_n \right).$$

We claim that
(196)
$$\min\left\{ \int_0^r a_k\left(\gamma_{0\ldots 0 u_k 0 \ldots 0}\left(P_k, t\right)\right) dt, \int_0^r a_k\left(\gamma_{0\ldots 0 u_k 0 \ldots 0}\left(P_{k+1}, t\right)\right) dt \right\} > \frac{\varepsilon}{4} B_k\left(r\right)$$

for $2 \leq k \leq n$ provided

(197)
$$2 C_L r \left|u_k\right| < 1 \quad \text{and} \quad 4 C_L r\left(n - 1\right) < \varepsilon.$$

To prove this, first note that since $a_k\left(x\right)$ is Lipschitz continuous in $x_2, \ldots, x_n$,

$$\int_0^r a_k\left(\gamma_{0\ldots 0 u_k 0 \ldots 0}\left(P_k, t\right)\right) dt =$$

$$\int_0^r a_k\left(q_1 + t, q_2, \ldots, q_{k-1}, p_k + u_k \int_0^t a_k\left(\gamma_{0\ldots 0 u_k 0 \ldots 0}\left(P_k, s\right)\right) ds, p_{k+1}, \ldots, p_n \right) dt$$

differs in absolute value from

$$\int_0^r a_k(q_1 + t, q_2, \ldots, q_{k-1}, p_k, p_{k+1}, \ldots, p_n)\, dt$$

by less than

$$r C_L \left|u_k\right| \int_0^r a_k\left(\gamma_{0\ldots 0 u_k 0 \ldots 0}\left(P_k, s\right)\right) ds \leq \frac{1}{2} \int_0^r a_k\left(\gamma_{0\ldots 0 u_k 0 \ldots 0}\left(P_k, t\right)\right) dt$$

provided $2 C_L r \left|u_k\right| < 1$. Hence, if $2 C_L r \left|u_k\right| < 1$, then
(198)
$$\int_0^r a_k\left(\gamma_{0\ldots 0 u_k 0 \ldots 0}\left(P_k, t\right)\right) dt > \frac{1}{2} \int_0^r a_k(q_1 + t, q_2, \ldots, q_{k-1}, p_k, p_{k+1}, \ldots, p_n)\, dt.$$

Next, due to the sizes of the edgelengths of $\widetilde{B}(0, r)$ and the fact that $a_k\left(x\right)$ is Lipschitz continuous in $x_2, \ldots, x_n$, the integral

$$\int_0^r a_k(q_1 + t, q_2, \ldots, q_{k-1}, p_k, p_{k+1}, \ldots, p_n)\, dt$$

differs in absolute value from

$$\int_0^r a_k(q_1 + t, q_2, \ldots, q_{k-1}, q_k, q_{k+1}, \ldots, q_n)\, dt = A_k(Q, r)$$

by at most

$$r C_L \left(2 B_k\left(r\right) + \cdots + 2 B_n\left(r\right)\right) \leq 2 r C_L(n - 1) B_k\left(r\right) \leq \frac{\varepsilon}{2} B_k\left(r\right)$$

due to the ordering (194) and the restrictions (197) on r. But $A_k(Q, r) \geq \varepsilon B_k(r)$ by (195) since $Q \in H = \mathcal{E}$. Hence,

$$\int_0^r a_k(q_1 + t, q_2, \ldots, q_{k-1}, p_k, p_{k+1}, \ldots, p_n) \, dt > \frac{\varepsilon}{2} B_k(r)$$

for such r, ε. The desired estimate for the first term in (196) follows by combining estimates. To estimate the second term, we replace P_k by P_{k+1} and p_k by q_k in the argument above. This completes the proof of (196) for $k = 2, \ldots, n$.

By doubling of a_k in x_1 there is a constant c with $0 < c < 1$ such that

$$\int_0^{C^{-1}r} a_k(q_1 + t, q_2, \ldots, q_{k-1}, p_k, p_{k+1}, \ldots, p_n) \, dt$$

$$\geq c \int_0^r a_k(q_1 + t, q_2, \ldots, q_{k-1}, p_k, p_{k+1}, \ldots, p_n) \, dt$$

where C is as in condition 1 of Definition 78. Combining the last two inequalities with $C^{-1}r \leq B_1(r)$ from condition 1 of Definition 78, we obtain

$$(199) \qquad \int_0^{B_1(r)} a_k(q_1 + t, q_2, \ldots, q_{k-1}, p_k, p_{k+1}, \ldots, p_n) \, dt > \frac{c\varepsilon}{2} B_k(r).$$

Hence combining (198) and (199), along with a similar argument for P_{k+1} in place of P_k, we obtain
(200)
$$\min\left\{ \int_0^{B_1(r)} a_k(\gamma_{0\ldots0u_k0\ldots0}(P_k, t)) \, dt, \int_0^{B_1(r)} a_k(\gamma_{0\ldots0u_k0\ldots0}(P_{k+1}, t)) \, dt \right\} > \frac{c\varepsilon}{4} B_k(r).$$

We now make a *fixed* choice of $u_k > 0$ which is independent of k, P, Q and the sequence $\{P_k\}$, namely let

$$u_k = \frac{8}{c\varepsilon}, \qquad 2 \leq k \leq n.$$

Then, multiplying (200) by u_k, we obtain

$$(201) \qquad \min\left\{ u_k \int_0^{B_1(r)} a_k(\gamma_{0\ldots0u_k0\ldots0}(P_k, s)) \, ds, \right.$$

$$\left. u_k \int_0^{B_1(r)} a_k(\gamma_{0\ldots0u_k0\ldots0}(P_{k+1}, s)) \, ds \right\}$$

$$> u_k \frac{c\varepsilon}{4} B_k(r) = 2B_k(r)$$

provided (see (197)) $16 C_L r(n-1) < c\varepsilon$, i.e., provided r is sufficiently small. Note that the term $2B_k(r)$ on the right side of (201) is the sidelength of $\widetilde{B}(0, r)$ in the direction of the x_k-axis. Thus (201) expresses the fact that by the time $t = B_1(0, r)$, both of the curves $\gamma_{0\ldots0u_k0\ldots0}(P_k, t)$ and $\gamma_{0\ldots0u_k0\ldots0}(P_{k+1}, t)$ have escaped out the " k^{th} side" of $\widetilde{B}(0, r)$ due to growth in their k^{th} coordinates.

The argument is now slightly different depending on which of p_k or q_k is smaller. Suppose first that $p_k < q_k$. With u_k chosen as above, (201) then implies that the curves

$$\gamma_{0\ldots0u_k0\ldots0}(P_k, t) \quad \text{and} \quad \gamma_0(P_{k+1}, t)$$

must intersect for a value $t = t_k$ depending on P_k with $0 < t_k < B_1(r)$:

$$\gamma_{0\ldots0u_k0\ldots0}(P_k, t_k) = \gamma_0(P_{k+1}, t_k) = (q_1 + t_k, q_2, \ldots, q_k, p_{k+1}, \ldots, p_n).$$

This follows from the forms of the two curves since the term $2B_k(r)$ in (201) is the kth sidelength of $\widetilde{B}(0, r)$. Recall that $u_k = \frac{8}{c\varepsilon}$ is positive and independent of k, P_k and P_{k+1}. In fact, the curves intersect for the value of t_k with

$$p_k + u_k \int_0^{t_k} a_k\left(\gamma_{0\ldots0u_k0\ldots0}(P_k, s)\right) ds = q_k.$$

Thus,

$$f\left(\gamma_{0\ldots0u_k0\ldots0}(P_k, t_k)\right) = f\left(\gamma_0(P_{k+1}, t_k)\right),$$

and therefore,

$$
\begin{aligned}
|f(P_k) - f(P_{k+1})| &\leq |f(P_k) - f\left(\gamma_{0\ldots0u_k0\ldots0}(P_k, t_k)\right)| \\
&\quad + |f\left(\gamma_0(P_{k+1}, t_k)\right) - f(P_{k+1})| \\
&= |f\left(\gamma_{0\ldots0u_k0\ldots0}(P_k, 0)\right) - f\left(\gamma_{0\ldots0u_k0\ldots0}(P_k, t_k)\right)| \\
&\quad + |f\left(\gamma_0(P_{k+1}, t_k)\right) - f\left(\gamma_0(P_{k+1}, 0)\right)| \\
&= \left|\int_0^{t_k} \frac{d}{dt} f\left(\gamma_{0\ldots0u_k0\ldots0}(P_k, t)\right) dt\right| \\
&\quad + \left|\int_0^{t_k} \frac{d}{dt} f\left(\gamma_0(P_{k+1}, t)\right) dt\right|,
\end{aligned}
$$

which is dominated by

$$\int_0^{t_k} \left|\frac{\partial f}{\partial x_1}\left(\gamma_{0\ldots0u_k0\ldots0}(P_k, t)\right)\right| dt$$

$$+ u_k \int_0^{t_k} \left|\frac{\partial f}{\partial x_k}\left(\gamma_{0\ldots0u_k0\ldots0}(P_k, t)\right) a_k\left(\gamma_{0\ldots0u_k0\ldots0}(P_k, t)\right)\right| dt$$

$$+ \int_0^{t_k} \left|\frac{\partial f}{\partial x_1}\left(\gamma_0(P_{k+1}, t)\right)\right| dt.$$

Assume for the moment that q lies in the left half of $\widetilde{B}(0, r)$, i.e. $q_1 \in [-B_1(0, r), 0]$. Then by construction and the fact that G and H are subsets of the rectangle $\widetilde{B}(0, r)$, the parts of both curves corresponding to t values with $0 \leq t \leq t_k$ lie in the set $\widetilde{B}(0, r)$. On the other hand, if q lies in the right half of $\widetilde{B}(0, r)$, i.e. $q_1 \in [0, B_1(0, r)]$, then replace the parameter t in the definition of the integral curves with $-t$, which we refer to as backward curves. Since (200) also holds with integration over the interval $[-B_1(0, r), 0]$ instead of $[0, B_1(0, r)]$, we conclude that the backward curves then lie in the set $\widetilde{B}(0, r)$ as well. Hence, letting $\chi = \chi_{\widetilde{B}(0,r)}$ denote the characteristic function of $\widetilde{B}(0, r)$, we obtain for $2 \leq k \leq n$,

$$(202) \qquad |f(P_k) - f(P_{k+1})|$$

$$\leq \left(1 + \frac{8}{c\varepsilon}\right) \int_0^{B_1(r)} |\nabla_a f\left(\gamma_{0\ldots0u_k0\ldots0}(P_k, t)\right)| \, \chi\left(\gamma_{0\ldots0u_k0\ldots0}(P_k, t)\right) dt$$

$$+ \int_0^{B_1(r)} |\nabla_a f\left(\gamma_0(P_{k+1}, t)\right)| \, \chi\left(\gamma_0(P_{k+1}, t)\right) dt.$$

If on the other hand $q_k < p_k$, then the integral curves

$$\gamma_0 (P_k, t) \quad \text{and} \quad \gamma_{0...0 u_k 0...0} (P_{k+1}, t)$$

intersect for some t_k with $0 < t_k < B_1 (0, r)$, namely for the value of t_k with

$$q_k + u_k \int_0^{t_k} a_k (\gamma_{0...0 u_k 0...0} (P_{k+1}, s)) \, ds = p_k,$$

where u_k is the same as before. Again, both curves lie in $\widetilde{B} (0, r)$ for $0 \leq t \leq t_k$ (by running the curve forward if q lies in the left half of $\widetilde{B} (0, r)$, and backward if q lies in the right half of $\widetilde{B} (0, r)$), and as above we obtain

$$(203) \qquad |f (P_k) - f (P_{k+1})| \leq \int_0^{B_1 (r)} |\nabla_a f (\gamma_0 (P_k, t))| \, \chi (\gamma_0 (P_k, t)) \, dt$$

$$+ \left(1 + \frac{8}{c\varepsilon}\right) \int_0^{B_1 (r)} |\nabla_a f (\gamma_{0...0 u_k 0...0} (P_{k+1}, t))| \, \chi (\gamma_{0...0 u_k 0...0} (P_{k+1}, t)) \, dt.$$

In any case, $|f (P_k) - f (P_{k+1})|$ is bounded by the sum of four terms, two from (202) and two from (203). Thus,

$$\left| \frac{1}{|G|} \int_G f (z) \, dz - \frac{1}{|H|} \int_H f (z) \, dz \right| \leq \frac{1}{|G| \, |H|} \int_H \int_G |f (P) - f (Q)| \, dQ \, dP$$

$$\leq \left(1 + \frac{8}{c\varepsilon}\right) \sum_{k=1}^n \frac{1}{|G| \, |H|} \int_H \int_G \left\{ \int_0^{B_1 (r)} E_k (t, P, Q) \, dt \right\} dQ \, dP,$$

where for $2 \leq k \leq n$,

$$(204) \; E_k (t, P, Q) \;\; = \;\; |\nabla_a f (\gamma_{0...0 u_k 0...0} (P_k, t))| \, \chi (\gamma_{0...0 u_k 0...0} (P_k, t))$$

$$+ |\nabla_a f (\gamma_0 (P_{k+1}, t))| \, \chi (\gamma_0 (P_{k+1}, t))$$

$$+ |\nabla_a f (\gamma_0 (P_k, t))| \, \chi (\gamma_0 (P_k, t))$$

$$+ |\nabla_a f (\gamma_{0...0 u_k 0...0} (P_{k+1}, t))| \, \chi (\gamma_{0...0 u_k 0...0} (P_{k+1}, t)),$$

and for $k = 1$,

$$(205) \qquad E_1 (t, P, Q) = |\nabla_a f (\gamma_0 (P, t))| \, \chi (\gamma_0 (P, t)).$$

Each of the corresponding four integral expressions in (204), as well as the corresponding integral expression in (205), can be estimated similarly, and we shall consider only the first one in (204) in detail.

For $2 \leq k \leq n$, we have

$$(206) \qquad \frac{1}{|G| \, |H|} \int_H \int_G \left\{ \int_0^{B_1 (r)} |\nabla_a f (\gamma_{0...0 u_k 0...0} (P_k, t))| \right.$$

$$\times \; \chi (\gamma_{0...0 u_k 0...0} (P_k, t)) \, dt \} \, dQ \, dP$$

$$\leq \quad \frac{1}{\alpha^2 \prod_{j=1}^n [2 B_j (r)]^2} \int_{-B_1 (r)}^{B_1 (r)} \cdots \int_{-B_n (r)}^{B_n (r)}$$

$$\times \int_{-B_1 (r)}^{B_1 (r)} \cdots \int_{-B_n (r)}^{B_n (r)} \int_0^{B_1 (r)} |\nabla_a f| \, \chi \, dt \, dq_n \ldots dq_1 dp_n \ldots dp_1,$$

where we have used the notation

$$|\nabla_a f| \, \chi =$$

$$\left| \nabla_a f \left(q_1 + t, q_2, \ldots, q_{k-1}, p_k + u_k \int_0^t a_k \left(\gamma_{0\ldots 0 u_k 0 \ldots 0} \left(P_k, s \right) \right) ds, p_{k+1}, \ldots p_n \right) \right| \chi(\ldots),$$

and $\chi(\ldots)$ denotes χ evaluated at the same point as $\nabla_a f$. Note that this expression is independent of $p_1, \ldots, p_{k-1}$ and $q_k, \ldots, q_n$. Recall that $u_k = \frac{8}{c\varepsilon}$ is a fixed constant independent of k. Performing the integration in $q_k, \ldots, q_n$ and $p_1, \ldots, p_{k-1}$ shows that (206) is at most

$$\frac{2^n}{2^{2n} \alpha^2 \prod_1^n B_j (r)^2} B_k (r) \cdots B_n (r) \, B_1 (r) \, B_2 (r) \cdots B_{k-1} (r) \int_{-B_k(r)}^{B_k(r)} \cdots \int_{-B_n(r)}^{B_n(r)}$$

$$\times \int_{-B_1(r)}^{B_1(r)} \cdots \int_{-B_{k-1}(r)}^{B_{k-1}(r)} \int_0^r |\nabla_a f| \, \chi \, dt \, dq_{k-1} \ldots dq_1 dp_n \ldots dp_k$$

(207)
$$\leq \frac{1}{\alpha^2 \prod_1^n 2B_j(r)} \int_{-B_1(r)}^{B_1(r)} \int_{\tilde{B}(0,r)} |\nabla_a f(w)|$$

$$\times \left| \frac{\partial(w_1, \ldots, w_n)}{\partial(t, q_2, \ldots, q_{k-1}, p_k, \ldots, p_n)} \right|^{-1} dw_1 \ldots dw_n dq_1,$$

where, for fixed q_1,

$$w = (w_1, \ldots, w_n)$$

$$= \left(q_1 + t, q_2, \ldots, q_{k-1}, p_k + u_k \int_0^t a_k \left(\gamma_{0\ldots 0 u_k 0 \ldots 0} \left(P_k, s \right) \right) ds, p_{k+1}, \ldots p_n \right).$$

We compute that

$$\frac{\partial(w_1, \ldots, w_n)}{\partial(t, q_2, \ldots, q_{k-1}, p_k, \ldots, p_n)}$$

is a determinant whose k^{th} row is the vector ∇w_k, whose diagonal entries other than the one in the kth row are 1, and whose remaining entries are all 0. Consequently, (208)

$$\frac{\partial(w_1, \ldots, w_n)}{\partial(t, q_2, \ldots, q_{k-1}, p_k, \ldots, p_n)} = \frac{\partial w_k}{\partial p_k} = 1 + \frac{\partial}{\partial p_k} \left(u_k \int_0^t a_k \left(\gamma_{0\ldots 0 u_k 0 \ldots 0} \left(P_k, s \right) \right) ds \right),$$

for $2 \leq k \leq n$.

We claim that for $2 \leq k \leq n$, the function

$$\Phi_k (x, t) = u_k \int_0^t a_k \left(\gamma_{0\ldots 0 u_k 0 \ldots 0} \left(x, s \right) \right) ds$$

is Lipschitz continuous in $x_2, \ldots, x_n$ with Lipschitz constant at most

$$C_L |u_k t| \left(1 - C_L |u_k t| \right)^{-1}.$$

We have from the fundamental theorem of calculus and the definition of integral curve that

$$\gamma_{0\ldots 0 u_k 0 \ldots 0} (x, s) = \Big(x_1 + s, x_2, \ldots, x_{k-1}, x_k$$

$$+ u_k \int_0^s a_k \left(\gamma_{0\ldots 0 u_k 0 \ldots 0} \left(x, \sigma \right) \right) d\sigma, x_{k+1}, \ldots, x_n \Big)$$

$$= (x_1 + s, x_2, \ldots, x_{k-1}, x_k + \Phi_k (x, s), x_{k+1}, \ldots, x_n).$$

Thus if $s > 0$, $x = (x_1, x_2, \ldots, x_n)$ and $x' = (x_1, x'_2, \ldots, x'_n)$, then since a_k is Lipschitz continuous in $x_2, \ldots, x_n$, we have

$$|u_k| \int_0^t |a_k (\gamma_{0 \ldots 0 u_k 0 \ldots 0} (x, s)) - a_k (\gamma_{0 \ldots 0 u_k 0 \ldots 0} (x', s))| \, ds$$

$$\leq |u_k| \int_0^t C_L \max \left\{ |x_2 - x'_2|, \ldots, |x_{k-1} - x'_{k-1}|, |\Phi_k (x, s) - \Phi_k (x', s)|, \right.$$

$$\left. |x_{k+1} - x'_{k+1}|, \ldots, |x_n - x'_n| \right\} ds$$

$$\leq |u_k| \int_0^t C_L \left\{ |x - x'| + |\Phi_k (x, s) - \Phi_k (x', s)| \right\} ds$$

$$\leq |u_k| \int_0^t C_L \left[|x - x'| + |u_k| \int_0^s |a_k (\gamma_{0 \ldots 0 u_k 0 \ldots 0} (x, \sigma)) \right.$$

$$\left. - a_k (\gamma_{0 \ldots 0 u_k 0 \ldots 0} (x', \sigma))| \, d\sigma \right] ds$$

$$\leq |u_k| C_L t \left[|x - x'| + |u_k| \int_0^t |a_k (\gamma_{0 \ldots 0 u_k 0 \ldots 0} (x, \sigma)) - a_k (\gamma_{0 \ldots 0 u_k 0 \ldots 0} (x', \sigma))| \, d\sigma \right].$$

Absorbing the second term on the right gives
(209)
$$|u_k| \int_0^t |a_k (\gamma_{0 \ldots 0 u_k 0 \ldots 0} (x, s)) - a_k (\gamma_{0 \ldots 0 u_k 0 \ldots 0}(x', s))| \, ds \leq \frac{|u_k| C_L t}{1 - |u_k| C_L t} |x - x'|,$$

which proves the claim since $|\Phi_k (x, t) - \Phi_k (x', t)|$ is bounded by the expression on the left side of (209). If $|u_k C_L t| \leq 1/3$, then $\frac{|u_k| C_L t}{1 - |u_k| C_L t} \leq \frac{1}{2}$, and thus the function $\Phi_k (x, t)$ has Lipschitz constant at most $\frac{1}{2}$ in the variables $x_2, \ldots, x_n$. It then follows from (208) that

(210)
$$\left| \frac{\partial(w_1, \ldots, w_n)}{\partial(t, q_2, \ldots, q_{k-1}, p_k, \ldots, p_n)} \right| \geq \frac{1}{2}.$$

The restriction $|u_k C_L t| \leq 1/3$ is satisfied by assuming that $24 C_L C r \leq c\varepsilon$, since $0 \leq t \leq B_1 (r) \leq C r$ by Definition 78 and $u_k = \frac{8}{c\varepsilon}$. Similar arguments apply to the other three terms in (204) and the single term in (205). For example, to treat the last term in (204), we replace p_k by q_k and P_k by P_{k+1} at the appropriate places above and perform the integrations in $q_{k+1}, \ldots, q_n$ and $p_1, p_2, \ldots, p_k$, with remaining integrations in $t, q_1, q_2, \ldots, q_k, p_{k+1}, \ldots, p_n$. The other two terms in (204), along with the term in (205), are simpler since u_k is replaced by 0 in them.

Using the estimate (210) on the right side of (207), and then integrating in q_1, we obtain that (206) is at most

$$\frac{2 B_1 (r)}{\alpha^2 \prod_{j=1}^n 2 B_j (r)} \int_{\tilde{B}(0, r)} |\nabla_a f (w)| \, 2 dw_1 \ldots dw_n.$$

Consequently, by combining estimates (recall that there are 4 terms in (204) and that we must sum over k for $k = 1, 2, \ldots, n$), we obtain

$$\left| \frac{1}{|G|} \int_G f (z) \, dz - \frac{1}{|H|} \int_H f (z) \, dz \right|$$

$$\leq \left(1 + \frac{8}{c\varepsilon} \right) \frac{8n}{\alpha^2} \frac{2 B_1 (r)}{\prod_{j=1}^n 2 B_j (r)} \int_{\tilde{B}(0, r)} |\nabla_a f (w)| \, dw.$$

This completes the proof of the basic Lemma 79 since $B_1(r) \approx r$ by Definition 78.

4.1. The proportional subrepresentation inequality. We can now obtain the proportional subrepresentation inequality (182) from Lemma 79 if we assume in addition that the family $\mathcal{B}$ is Ω-locally equivalent to the family $\widetilde{\mathcal{B}}$ in the sense of subsection 2.3, i.e.

$$(211) \qquad B\left(x, C^{-1}r\right) \subset \widetilde{B}(x,r) \subset B(x, Cr), \qquad x \in \Omega, \, 0 < r < \delta \, dist\,(x, \partial\Omega).$$

Note that we actually have $B(x,r) \subset \widetilde{B}(x,r)$ by definition of $\widetilde{B}(x,r)$.

PROPOSITION 80. *Assume that a_i is nonnegative and continuous for $x \in \Omega$, doubling in x_1 uniformly in $x_2, \ldots, x_n$ and Lipschitz continuous in $x_2, \ldots, x_n$ uniformly in x_1. Let $X_1 = \frac{\partial}{\partial x_1}$, $X_2(x) = a_2(x)\frac{\partial}{\partial x_2}$, $\ldots$, $X_n(x) = a_n(x)\frac{\partial}{\partial x_n}$ and suppose that the collection of vector fields $\mathcal{X} = \{X_j(x)\}_{j=1}^n$ is adapted on Ω to a prehomogeneous space $\mathcal{B}$ on Ω, and that $\mathcal{B}$ is Ω-locally equivalent to the family $\widetilde{\mathcal{B}}$ in (89). Then the proportional subrepresentation inequality (182) of Axiom 1 holds.*

PROOF. Since the families $\mathcal{B}$ and $\widetilde{\mathcal{B}}$ are Ω-locally equivalent, there is $K \geq 1$ such that

$$(212) \qquad \widetilde{B}(x,r) \subset B(x, Kr), \qquad 0 < r < \delta \, dist\,(x, \partial\Omega).$$

Recall that the rectangles $\widetilde{B}(x,r)$ are given by

$$\widetilde{B}(x,r) = I_1(x,r) \times \prod_{k=2}^n [x_k - B_k(x,r), x_k + B_k(x,r)],$$

where $I_1(x,r)$ is the interval

$$I_1(x,r) = [x_1 - B_1(x,r), x_1 + B_1(x,r)].$$

From condition 1 of the assumption that the vector fields $\mathcal{X} = \{X_j(x)\}_{j=1}^n$ are adapted to $\mathcal{B}$, we obtain for some $\alpha > 0$,

$$(213) \qquad \alpha < \frac{B_1(x,r)}{r} < \frac{1}{\alpha}, \qquad 0 < r < \delta \, dist\,(x, \partial\Omega).$$

We decrease α if necessary to achieve $\alpha \leq c$ where c is the constant in the weak monotonicity condition (60).

We first prove (182) with $C = 1$ for $x \in B(y, R)$ in the special case $x = y$, and we will further assume without loss of generality that $x = y = 0$ and $0 < R < \delta \, dist\,(0, \partial\Omega)$ is fixed. Let $\tau \in (0,1)$ to be fixed near the end of the proof. Let $\tau_m = \tau^m R$ for $m = 1, 2, \ldots$ and define the sequence of points $y^m = (\tau_m, 0, \ldots, 0)$, $m \geq 1$. Note that y^m tends to $0 = (0, \ldots, 0)$ as $m \to \infty$. Let γ be the constant in the engulfing condition (56). Consider the sets, which we refer to as preballs,

$$B_m = B\left(y^m, \alpha(1-\tau)\tau_m\right), B_m^* = B\left(y^m, K\alpha^{-1}\gamma(1-\tau)\tau_m\right),$$

and their corresponding smallest closed containing rectangles

$$\widetilde{B_m} = \widetilde{B}\left(y^m, \alpha(1-\tau)\tau_m\right), \widetilde{B_m^*} = \widetilde{B}\left(y^m, K\alpha^{-1}\gamma(1-\tau)\tau_m\right),$$

for $m \geq 1$. We claim the following two properties hold with $\tau \geq 1 - \frac{\alpha^2}{2K\gamma}$:

$$(214) \qquad \begin{cases} \widetilde{B_m} \cup \widetilde{B_{m+1}} \subset \widetilde{B_m^*}, & m \geq 1 \\ \sum_{m=M}^{\infty} \chi_{\widetilde{B_m^*}} \leq C\chi_{\widetilde{B}(0, C'\tau^M R)}, & M \geq 1 \end{cases},$$

with $C = C(\tau)$ and $C' = \frac{K^2\gamma^2}{\alpha}$.

Now $\widetilde{B_m} \subset \widetilde{B_m^*}$ by weak monotonicity since $\alpha \leq c < 1$ and $K, \gamma \geq 1$ imply that $B_m \subset B_m^*$. To see that $\widetilde{B_{m+1}} \subset \widetilde{B_m^*}$, we first note that by (213) with r replaced by $\alpha^{-1}(1-\tau)\tau_m$,

$$\tau_{m+1} = \tau\tau_m \in [\tau_m - (1-\tau)\tau_m, \tau_m + (1-\tau)\tau_m] \subset I_1\left(y^m, \alpha^{-1}(1-\tau)\tau_m\right)$$

and so $y^{m+1} \in \widetilde{B}\left(y^m, \alpha^{-1}(1-\tau)\tau_m\right)$. From (212) we then obtain

$$y^{m+1} \in B\left(y^m, K\alpha^{-1}(1-\tau)\tau_m\right).$$

Thus the engulfing property (56) of the preballs yields

$$B\left(y^{m+1}, K\alpha^{-1}(1-\tau)\tau_m\right) \subset B\left(y^m, \gamma K\alpha^{-1}(1-\tau)\tau_m\right) = B_m^*.$$

Using $\alpha \leq c < 1$ and the weak monotonicity property (60) we have

$$B_{m+1} = B\left(y^{m+1}, \alpha(1-\tau)\tau_{m+1}\right) \subset B\left(y^{m+1}, K\alpha^{-1}(1-\tau)\tau_m\right).$$

Combining these containments, we obtain $B_{m+1} \subset B_m^*$ and so $\widetilde{B_{m+1}} \subset \widetilde{B_m^*}$, thus proving the first assertion in (214).

For the remaining assertion in (214), we first claim that the intervals

$$\left\{I_1\left(y^m, K\alpha^{-1}\gamma(1-\tau)\tau_m\right)\right\}_{m=1}^{\infty}$$

have finite overlap N where $N = N(\tau)$. Indeed, from (213) we have

$$I_1\left(y^{m+N}, K\alpha^{-1}\gamma(1-\tau)\tau_{m+N}\right)$$
$$\subset \left[\tau_{m+N} - K\alpha^{-2}\gamma(1-\tau)\tau_{m+N}, \tau_{m+N} + K\alpha^{-2}\gamma(1-\tau)\tau_{m+N}\right].$$

Thus we will have

$$I_1\left(y^{m+N}, K\alpha^{-1}\gamma(1-\tau)\tau_{m+N}\right) \cap I_1\left(y^m, K\alpha^{-1}\gamma(1-\tau)\tau_m\right) = \phi$$

for all $m \geq 1$ if

$$\tau_{m+N} + K\alpha^{-2}\gamma(1-\tau)\tau_{m+N} < \tau_m - K\alpha^{-2}\gamma(1-\tau)\tau_m, \quad m \geq 1,$$

which in turn holds provided

$$\tau^N\left(1 + K\alpha^{-2}\gamma(1-\tau)\right) < 1 - K\alpha^{-2}\gamma(1-\tau).$$

This latter inequality can be achieved by choosing $\tau \geq 1 - \frac{\alpha^2}{2K\gamma}$ and then N so large that $\tau^N \frac{3}{2} < \frac{1}{2}$, i.e.,

$$N > \frac{\log(1/3)}{\log\tau},$$

proving our claim about the intervals I_1.

Second, we observe that

$$\tau_m = \tau^m R \in [-K\gamma\tau^m R, K\gamma\tau^m R] \subset I_1\left(0, K\alpha^{-1}\gamma\tau^m R\right)$$

shows that $y^m \in \widetilde{B}\left(0, K\alpha^{-1}\gamma\tau^m R\right)$. Thus by (212), $y^m \in B\left(0, K^2\alpha^{-1}\gamma\tau^m R\right)$, and then by engulfing, $B\left(y^m, K^2\alpha^{-1}\gamma\tau_m\right) \subset B\left(0, K^2\alpha^{-1}\gamma^2\tau^m R\right)$ (note the square on the last γ). Hence

$$\widetilde{B_m^*} = \widetilde{B}\left(y^m, K\alpha^{-1}\gamma(1-\tau)\tau_m\right) \subset \widetilde{B}\left(0, K\alpha^{-1}\gamma^2\tau^m R\right), \quad m \geq 1,$$

and this completes the proof of (214).

Now let f be Lipschitz continuous on $\widetilde{B}\left(0, C'R\right)$ and vanish on a subset $\mathcal{E}$ of $B\left(0, R\right)$ satisfying $\left|\mathcal{E}\right| \geq \beta\left|B\left(0, R\right)\right|$. By the continuity of f at 0 we have

$$\left|f\left(0\right)\right| = \left|\lim_{m \to \infty} \frac{1}{\left|\widetilde{B_m}\right|} \int_{\widetilde{B_m}} f\left(z\right)\, dz\right|$$

$$\leq \left|\frac{1}{\left|\widetilde{B_M}\right|} \int_{\widetilde{B_M}} f\left(z\right)\, dz\right| + \sum_{m=M}^{\infty} \left|\frac{1}{\left|\widetilde{B_{m+1}}\right|} \int_{\widetilde{B_{m+1}}} f\left(z\right)\, dz - \frac{1}{\left|\widetilde{B_m}\right|} \int_{\widetilde{B_m}} f\left(z\right)\, dz\right|$$

for all $M \geq 1$. Let M be the least positive integer such that $K^2 \alpha^{-1} \gamma^2 \tau^M \leq 1$, thus ensuring that $\widetilde{B_m^*} \subset \widetilde{B}\left(0, R\right)$ for all $m \geq M$. It is easy to see that there is then a positive constant δ such that $\widetilde{B_M}$ is a δ-proportional subset of $\widetilde{B}\left(0, R\right)$. Indeed, $\widetilde{B_M} \subset \widetilde{B_M^*}$ by (214) and then $\widetilde{B_M^*} \subset \widetilde{B}\left(0, R\right)$ by the choice of M as indicated above. The proportional assertion follows from $B_M \subset \widetilde{B_M}$, (212) and the fact that the "radius" of B_M is comparable to R, as M has been fixed. Also, there is a constant $c > 0$ such that $\mathcal{E}$ is a $c\beta$-proportional subset of $\widetilde{B}\left(0, R\right)$. Now apply Lemma 79 to the preball $B\left(0, R\right)$ with $\alpha = \min\left\{c\beta, \delta\right\}$ and $G = \widetilde{B_M}, H = \mathcal{E}$ to obtain

$$\left|\frac{1}{\left|\widetilde{B_M}\right|} \int_{\widetilde{B_M}} f\left(z\right)\, dz\right| = \left|\frac{1}{\left|\widetilde{B_M}\right|} \int_{\widetilde{B_M}} f\left(z\right)\, dz - \frac{1}{\left|\mathcal{E}\right|} \int_{\mathcal{E}} f\left(z\right)\, dz\right|$$

$$\leq C \frac{R}{\left|\widetilde{B}\left(0, R\right)\right|} \int_{\widetilde{B}\left(0,R\right)} \left|\nabla_a f\left(z\right)\right|\, dz.$$

It is again easy to see that there is a positive constant δ such that both $\widetilde{B_m}$ and $\widetilde{B_{m+1}}$ are δ-proportional subsets of $\widetilde{B_m^*}$ for all $m \geq 1$. Using (214) we can apply Lemma 79 to the preball B_m^* with $\alpha = \delta$ and $G = \widetilde{B_{m+1}}, H = \widetilde{B_m}$ to obtain

$$\left|\frac{1}{\left|\widetilde{B_{m+1}}\right|} \int_{\widetilde{B_{m+1}}} f\left(z\right)\, dz - \frac{1}{\left|\widetilde{B_m}\right|} \int_{\widetilde{B_m}} f\left(z\right)\, dz\right| \leq C \frac{\gamma K \alpha^{-1}\left(1 - \tau\right)\tau_m}{\left|\widetilde{B_m^*}\right|} \int_{\widetilde{B_m^*}} \left|\nabla_a f\left(z\right)\right|\, dz,$$

for all $m \geq 1$. Adding these estimates yields

$$\left|f\left(0\right)\right| \leq C \int \left|\nabla_a f\left(z\right)\right| \left\{\sum_{m=M}^{\infty} \frac{\tau_m}{\left|\widetilde{B_m^*}\right|} \chi_{\widetilde{B_m^*}}\left(z\right)\right\}\, dz + C \frac{R}{\left|\widetilde{B}\left(0, R\right)\right|} \int_{\widetilde{B}\left(0,R\right)} \left|\nabla_a f\left(z\right)\right|\, dz.$$

We now claim that

$$\sum_{m=M}^{\infty} \frac{\tau_m}{\left|\widetilde{B_m^*}\right|} \chi_{\widetilde{B_m^*}}\left(z\right) \leq C \chi_{\widetilde{B}\left(0,R\right)}\left(z\right) \frac{d\left(0, z\right)}{\left|B\left(0, d\left(0, z\right)\right)\right|}, \quad z \in \widetilde{B}\left(0, R\right),$$

where $d\left(0, z\right) = \inf\left\{r > 0 : z \in B\left(0, r\right)\right\}$. Indeed, we will momentarily check using engulfing and (212) that $d\left(0, z\right) \approx \tau_m$ for $z \in \widetilde{B_m^*}$, and that $\left|\widetilde{B_m^*}\right| \approx \left|\widetilde{B}\left(0, \tau_m\right)\right|$, $m \geq 1$. With this done, and since there are at most N rectangles $\widetilde{B_m^*}$ that contain a fixed point z, and since $\widetilde{B_m^*} \subset \widetilde{B}(0,R)$ for $m \geq M$, we then have

$$\sum_{m=M}^{\infty} \frac{\tau_m}{\left|\widetilde{B_m^*}\right|} \chi_{\widetilde{B_m^*}}(z) \leq CN\chi_{\widetilde{B}(0,R)}(z) \frac{d(0,z)}{\left|\widetilde{B}(0,d(0,z))\right|}$$

as required.

So it remains to show $d(0,z) \approx \tau_m$ for $z \in \widetilde{B_m^*}$, and $\left|\widetilde{B_m^*}\right| \approx \left|\widetilde{B}(0,\tau_m)\right|$, $m \geq 1$. Now $\tau_m \approx B_1(0,\tau_m)$ by condition 1 of Definition 78, and so there is a constant $C \geq 1$ such that

$$y^m = (\tau_m,0,...,0) \in \widetilde{B}(0,C\tau_m) \setminus \widetilde{B}\left(0,\frac{1}{C}\tau_m\right).$$

By (212), we then have for a larger constant C',

$$y^m \in B(0,C'\tau_m) \setminus B\left(0,\frac{1}{C'}\tau_m\right),$$

which establishes that $d(0,y^m) \approx \tau_m$ with constant of equivalence independent of $m \geq 1$. Now take $z \in \widetilde{B_m^*}$. Now we recall Remark 35 which shows that d is a quasimetric. Then by what we just proved and (54), we have

$$\begin{aligned}
\tau_m &\leq Cd(0,y^m) \leq C\kappa[d(0,z)+d(y^m,z)] \\
&\leq C\kappa[d(0,z)+C(1-\tau)\tau_m].
\end{aligned}$$

If we now choose $1-\tau$ sufficiently small, we can absorb the second term on the right side to obtain that $\tau_m \leq Cd(0,z)$. Conversely, we first recall from Remark 36 that setting $z = x$ in (54) yields $d(x,y) \leq \kappa d(y,x)$. Together with $d(y^m,z) \leq C(1-\tau)\tau_m$ proved above, we thus have

$$\begin{aligned}
d(0,z) &\leq \kappa[d(0,y^m)+d(z,y^m)] \\
&\leq \kappa[d(0,y^m)+\kappa d(y^m,z)] \\
&\leq \kappa[C\tau_m+\kappa C(1-\tau)\tau_m] \\
&\leq C'\tau_m,
\end{aligned}$$

and this completes our first assertion that $d(0,z) \approx \tau_m$ for $z \in \widetilde{B_m^*}$.

Next let us show that $\left|\widetilde{B_m^*}\right| \approx \left|\widetilde{B}(0,\tau_m)\right|$ for $m \geq 1$. It is enough to show that $|B_m^*| \approx |B(0,\tau_m)|$ because of doubling and (212). If $\xi \in B(0,\tau_m)$, then $d(0,\xi) < \tau_m$ and so

$$\begin{aligned}
d(y^m,\xi) &\leq \kappa[d(y^m,0)+d(\xi,0)] \\
&\leq \kappa^2[d(0,y^m)+d(0,\xi)] \\
&\leq C\tau_m,
\end{aligned}$$

by what we have already shown. Since τ is now fixed, it follows that ξ belongs to a fixed enlargement B_m^{**} of B_m^*. Thus $B(0,\tau_m) \subset B_m^{**}$, and we obtain $|B(0,\tau_m)| \leq C|B_m^*|$ by doubling. The proof of the opposite inequality is similar.

This completes the proof of (182) with $C = 1$ in the case $x = y$. From this, together with the engulfing property of the preballs, it follows immediately that (182) holds for arbitrary $x \in B(y,R)$ with a large blowup constant C, and the proof of Proposition 80 is now complete.

REMARK 81. *Using the fact that the flag and noninterference balls are rect-angles, it is possible to show that in these cases, the proportional subresentation formula (182) holds with the blowup constant $C = 1$. The idea is to partition the rectangle into 2^n quadrants, and then, depending on which quadrant the point x lies in, to choose the shooting parameters u so that the integral curves emanating from x always remain in the rectangle. The details are left to the interested reader. See also* [27] *where geodesics are used to obtain the standard subrepresentation inequality (186) with blowup constant $C_0 = 1$ in certain special cases. We conjecture that in Proposition 80, we can in similar fashion obtain the proportional subrepresentation inequality (182) with blowup constant $C = 1$ by taking the centers y^m of sufficiently small preballs B_m in the above argument to lie on an appropropriate integral curve through x.*

We can now state and prove a general theorem on subellipticity of diagonal Lipschitz vector fields. Axiom 1 and Lemma 71 play a key role in the proof.

THEOREM 82. *Assume that a_i is nonnegative and continuous for $x \in \Omega$, doubling in x_1 uniformly in $x_2, \ldots, x_n$ and Lipschitz continuous in $x_2, \ldots, x_n$ uniformly in x_1. Let $X_1 = \frac{\partial}{\partial x_1}$, $X_2(x) = a_2(x) \frac{\partial}{\partial x_2}$, $\ldots$, $X_n(x) = a_n(x) \frac{\partial}{\partial x_n}$ and suppose that the collection of vector fields $\mathcal{X} = \{X_j(x)\}_{j=1}^n$ is adapted on Ω to a prehomogeneous space $\mathbb{B} = (\Omega, \mathcal{B})$ on Ω, and $\mathcal{B}$ that is Ω-locally equivalent to the family $\widetilde{\mathcal{B}}$ as in (211). Let D be a doubling exponent for the family $\mathcal{B}$. Finally we either suppose that the "accumulating sequence of Lipschitz cutoff functions" condition (20) holds for some $p > \max\{D, 4\}$, or we suppose that $\mathcal{B}$ is Ω-locally equivalent to the family of subunit balls $\mathcal{K}$. Then $\mathcal{X}$ is L^q-subelliptic in Ω for all $q > D$.*

PROOF. Proposition 80 shows that the proportional subrepresentation inequality (182) of Axiom 1 holds for the prehomogeneous space $\mathbb{B}$. Conditions (179) and (180) of Axiom 1 hold since $\mathcal{B}$ and $\widetilde{\mathcal{B}}$ are Ω-locally equivalent. Thus Axiom 1 holds for $\mathcal{X}$ and the prehomogeneous space $\mathbb{B}$. Lemma 71 then shows that the corresponding symmetric general homogeneous space $\mathbb{B}^* = (\Omega, \mathcal{B}^*)$ where $\mathcal{B}^* = \{B^*(x, r)\}_{x \in \Omega, 0 < r < \infty}$, also satisfies Axiom 1 relative to $\mathcal{X}$. We will now verify the hypotheses of Theorem 8 applied to the general homogeneous space $\mathbb{B}^*$ and the quadratic form $\mathcal{Q}(x, \xi) = \sum_{j=1}^n (a_j(x)\xi_j)^2$ arising from the collection of vector fields $\mathcal{X}$ under the assumption that (20) holds for some $p > \max\{D, 4\}$. The doubling condition (13) is obvious. The convex hull equivalence (179) from Axiom 1 together with Remark 5 then shows that the containment condition (11) holds. Proposition 74 and Lemma 72 show that the Sobolev and Poincaré inequalities (15) and (17) hold. The "accumulating sequence of Lipschitz cutoff functions" condition (20) is obviously inherited by equivalent families, as is a doubling exponent, and Theorem 8 now applies to complete the proof of Theorem 82 under the assumption that (20) holds for some $p > \max\{D, 4\}$ since

$$Q = \max\{Q^*, 2\sigma'\} = \max\{Q^*, D\} = D.$$

We appeal instead to Theorem 10 in the case that $\mathcal{B}$ is Ω-locally equivalent to $\mathcal{K}$, since then Axiom 1 holds for $\mathcal{X}$ and $\mathcal{K}$, the pair $(\Omega, \mathcal{K})$ is a symmetric general homogeneous space, and the hypotheses of Theorem 10 are seen to hold just as above.

5. The reduction of the proofs

Using Theorem 82 and Proposition 44, we now reduce the proofs of Theorems 17 and 20 to proving that the flag and noninterference balls satisfy the size-limiting condition (91) and form a δ-local prehomogeneous space and a δ-local homogeneous space respectively that is adapted to the collection of vector fields $\mathcal{X} = \{X_j\}_{j=1}^n$ as in Definition 78. As noted at the beginning of subsection 1.2.1 of the introduction, we may assume, without loss of generality, that $a_1 \equiv 1$ so that $X_1 = \frac{\partial}{\partial x_1}$.

We begin with the proof of Theorem 17, assuming in addition that the flag balls $\mathcal{B}$ satisfy (91) and form a δ-local prehomogeneous space $\mathbb{B} = (\Omega, \mathcal{B})$ adapted to $\mathcal{X}$. Since the conclusion of Theorem 17 is local in nature, we may fix $x_0 \in \Omega$ and restrict attention to the δ_0-local prehomogeneous space $\mathbb{B}_0$ that is induced on Ω_0 by $\mathbb{B}$ as in Proposition 41. Since $\mathbb{B}_0$ is extendible, and continues to be adapted to $\mathcal{X}$, we may as well assume that $\mathbb{B} = (\Omega, \mathcal{B})$ itself is a prehomogeneous space adapted to $\mathcal{X}$.

By Proposition 44, the flag balls are Ω-locally equivalent to the subunit balls $K(x,r)$. Indeed, condition (91) has been assumed and condition (90) of Proposition 44 holds with $C_{y,r} = f_{B(y,r)}$ by Lemma 72 and Proposition 80. The reverse Hölder and Lipschitz requirements on the vector fields in Proposition 44 follow from the hypotheses of Theorem 17. The flag balls are rectangles and so $\mathcal{B}$ is Ω-locally equivalent to $\widetilde{\mathcal{B}}$, and the remaining hypotheses of Theorem 82 are explicitly assumed in Theorem 17. Thus Theorem 82 applies to complete the proof of Theorem 17, assuming the flag balls $\mathcal{B}$ satisfy (91) and form a δ-local prehomogeneous space $\mathbb{B} = (\Omega, \mathcal{B})$ adapted to $\mathcal{X}$.

We turn now to the proof of Theorem 20, assuming in addition that the noninterference balls $\mathcal{A}$ satisfy (91) and form a δ-local homogeneous space $\mathbb{A} = (\Omega, \mathcal{A})$ adapted to $\mathcal{X}$. Again, Proposition 41 and Proposition 44 apply, and we conclude that we may assume that $\mathbb{A}$ itself is a homogeneous space adapted to $\mathcal{X}$, and that the noninterference balls $A(x,r)$ are Ω-locally equivalent to the subunit balls $K(x,r)$. For this, recall that the hypotheses of Theorem 20 include the continuity of the a_j, as well as the reverse Hölder condition in the x_1 variable, uniformly in the other variables. The noninterference balls are rectangles and so $\mathcal{A}$ is Ω-locally equivalent to $\widetilde{\mathcal{A}}$, and the remaining hypotheses of Theorem 82 are explicitly assumed in Theorem 20. Thus Theorem 82 applies to complete the proof of Theorem 20, assuming the noninterference balls $\mathcal{A}$ satisfy (91) and form a δ-local homogeneous space $\mathbb{A} = (\Omega, \mathcal{A})$ adapted to $\mathcal{X}$.

5.1. The sharper technical case. In this subsection, we prove the sharper sufficient conditions for subellipticity of vector fields in Theorems 23 and 24 under the additional assumption that the flag and noninterference balls form a δ-local prehomogeneous and homogeneous space respectively adapted to $\mathcal{X}$. We suppose as well that the noninterference balls $\mathcal{A}$ satisfy the size-limiting condition (91), and that the flag balls $\mathcal{B}$ satisfy the reverse doubling property (215). As above, Proposition 41 shows that we may assume that the flag and noninterference balls form a prehomogeneous and homogeneous space respectively. We will not demonstrate the equivalence of our family of balls with the subunit balls and in fact, if the coefficients a_j fail to be reverse Hölder of infinite order in the x_1 variable, then the noninterference balls $\mathcal{A}$ are *not* equivalent with the subunit balls $\mathcal{K}$ by Proposition 104. This requires that we establish the "accumulating sequence of Lipschitz cutoff functions" condition (20) directly for the flag and noninterference balls, assuming that the a_j satisfy a reverse Hölder condition of order $p < \infty$ in the x_1 variable. Note that now $\mathcal{Q}(x,\xi) = \sum_{j=1}^{n} a_j(x)^2 \xi_j^2$ with $a_1(x) = 1$, so that $\|\nabla\psi\|_\mathcal{Q} = |\nabla_a\psi|$ for any Lipschitz continuous function ψ. We first consider the noninterference balls $\mathcal{A}$.

PROPOSITION 83. *Let* $1 < p \leq \infty$. *The "accumulating sequence of Lipschitz cutoff functions" condition (20) holds for the* δ-*local homogeneous space* $\mathcal{A}$ *of non-interference balls* $A(x,r)$, *provided both the nondegeneracy condition (40) and size-limiting condition (91) with* $\mathcal{B} = \mathcal{A}$ *hold, and provided the* a_j *are reverse Hölder of order* p *in* x_1, *uniformly in* $x_2, ... x_n$, *and Lipschitz continuous in* $x_2, ... x_n$, *uniformly in* x_1.

PROOF. Fix a ball $A(y,r)$ with $y \in \Omega$, $0 < r < \delta\, dist(y,\partial\Omega)$, and let $A_j(t) = A_j(y,t)$. Note that $A_j(t)$ is strictly increasing as a function of t. Given $\frac{r}{2} < s < t \leq r$, define

$$\psi(x) = \prod_{j=1}^{n} \varphi\left(\frac{A_j(t) - |x_j - y_j|}{A_j(t) - A_j(s)}\right),$$

where

$$\varphi(\xi) = \begin{cases} 0, & \xi \leq 0 \\ \xi, & 0 \leq \xi \leq 1 \\ 1, & 1 \leq \xi \end{cases}.$$

We compute by the chain rule and the support conditions on φ that

$$|\nabla_a\psi(x)| \leq \sum_{j=1}^{n} a_j(x)\, \chi_{\{x \in A(y,r): A_j(s) \leq |x_j - y_j| \leq A_j(t)\}} \frac{1}{A_j(t) - A_j(s)},$$

and thus by Minkowski's inequality that

$$\left(\frac{1}{|A(y,r)|}\int |\nabla_a\psi|^p\right)^{\frac{1}{p}}$$

$$\leq \sum_{j=1}^{n}\left\{\frac{1}{|A(y,r)|}\int_{\{x \in A(y,r): A_j(s) \leq |x_j - y_j| \leq A_j(t)\}} a_j(x)^p \left(\frac{1}{A_j(t) - A_j(s)}\right)^p dx\right\}^{\frac{1}{p}}.$$

Fix $x' \in \prod_{i=2}^{n}(y_i - A_i(r), y_i + A_i(r))$ for the moment. Since a_j is reverse Hölder of order p in the first variable x_1, we have for $j \geq 2$,

$$\left(\frac{1}{2r}\int_{y_1-r}^{y_1+r} a_j(x_1,x')^p\, dx_1\right)^{\frac{1}{p}} \leq \frac{C}{r}\int_{y_1-r}^{y_1+r} a_j(x_1,x')\, dx_1 \leq \frac{C}{r}A_j((y_1,x'),r) \leq \frac{C}{r}A_j(r),$$

by the size-limiting condition (91) with $\mathcal{B} = \mathcal{A}$. Thus for $j \geq 2$,

$$\left\{ \frac{1}{|A(y,r)|} \int_{\{x \in A(y,r) : A_j(s) \leq |x_j - y_j| \leq A_j(t)\}} a_j(x)^p \, dx \right\}^{\frac{1}{p}}$$

$$= \left\{ \frac{1}{\prod_{i=2}^n 2A_i(r)} \int_{\prod_{i=2}^n (y_i - A_i(r), y_i + A_i(r))} \chi_{\{x_j : A_j(s) \leq |x_j - y_j| \leq A_j(t)\}} \right.$$

$$\left. \times \left(\frac{1}{2r} \int_{y_1 - r}^{y_1 + r} a_j(x_1, x')^p \, dx_1 \right) dx' \right\}^{\frac{1}{p}}$$

$$\leq \left\{ \frac{1}{\prod_{i=2}^n 2A_i(r)} 2(A_j(t) - A_j(s)) \left(\prod_{i \geq 2, i \neq j} 2A_i(r) \right) \left(\frac{C}{r} A_j(r) \right)^p \right\}^{\frac{1}{p}}$$

$$\leq \frac{C}{r} A_j(r) \left(\frac{A_j(t) - A_j(s)}{A_j(r)} \right)^{\frac{1}{p}}.$$

A similar estimate holds for the case $j = 1$, and since $A_j(r) \approx A_j(t)$ by doubling in the first variable (which is a standard consequence of the reverse Hölder assumption), we have

$$\left(\frac{1}{|A(y,r)|} \int |\nabla_a \psi|^p \right)^{\frac{1}{p}} \leq \frac{C}{r} \sum_{j=1}^n \left\{ \frac{A_j(t)}{A_j(t) - A_j(s)} \right\}^{1 - \frac{1}{p}}$$

$$= \frac{C}{r} \sum_{j=1}^n \left\{ \frac{\int_0^t a_j(y_1 + \xi, y') \, d\xi}{\int_s^t a_j(y_1 + \xi, y') \, d\xi} \right\}^{\frac{1}{p'}}$$

$$\leq \frac{C}{r} \left\{ \frac{t}{t - s} \right\}^{\frac{d}{p'}},$$

by doubling in the first variable again, where d is the doubling exponent in

$$\int_J a_j(x) \, dx_1 \leq C \left(\frac{|J|}{|I|} \right)^d \int_I a_j(x) \, dx_1, \quad \text{whenever } I \subset J, \ 2 \leq j \leq n \text{ uniformly}$$

$$\text{in } x_2, \ldots, x_n.$$

Now let $r_j - r_{j+1} = \frac{r}{5(j+1)^2}$, $r_0 = r$, $s = r_{j+1}$, $t = r_j$ and write ψ_j for ψ to obtain

$$\left(\frac{1}{|A(y,r)|} \int |\nabla_a \psi_j|^p \right)^{\frac{1}{p}} \leq \frac{C}{r} \left\{ \frac{r_j}{r_j - r_{j+1}} \right\}^{\frac{d}{p'}} \leq \frac{C}{r} (j+1)^{\frac{2d}{p'}}.$$

Note that $\lim_{j \to \infty} r_j = \inf r_j = cr$ for some c in $(\frac{1}{2}, 1)$. It is now easy to verify that the sequence $\{\psi_{2j}\}_{j=1}^\infty$ satisfies the conditions in the "accumulating sequence of Lipschitz cutoff functions" condition (20) with $N = \frac{2d}{p'}$. This completes the proof of Proposition 83.

We can now use the noninterference balls $A(x,r)$ in Theorem 82 to obtain Theorem 24, assuming the noninterference balls $\mathcal{A}$ satisfy the size-limiting condition (91) and form a δ-local homogeneous space $\mathbb{A} = (\Omega, \mathcal{A})$ adapted to $\mathcal{X}$. Indeed, we just finished establishing (20) under these conditions, and we observed earlier that Proposition 41 shows we may assume $\mathbb{A}$ itself is a homogeneous space. The remaining hypotheses of Theorem 82 are explicitly contained in Theorem 24. This completes the reduction of the proof of Theorem 24 to demonstrating that the noninterference balls $\mathcal{A}$ satisfy (91) and form a δ-local homogeneous space $\mathbb{A} = (\Omega, \mathcal{A})$ adapted to $\mathcal{X}$.

The reduction of the proof of Theorem 23 proceeds along similar lines. We must first establish condition (20) directly in the spirit of Proposition 83.

PROPOSITION 84. *Let* $1 < p \leq \infty$. *The "accumulating sequence of Lipschitz cutoff functions" condition (20) holds for the δ-local homogneous space $\mathcal{B}$ of flag balls $B(y,r)$, provided Definition 13 holds, the a_j are reverse Hölder of order p in x_1, uniformly in $x_2, ... x_n$, and Lipschitz continuous in $x_2, ... x_n$, uniformly in x_1, and finally provided there is a positive constant c_0 such that the following reverse doubling inequality holds:*

$$(215) \qquad B_j(y, c_0 r) \leq \frac{1}{2} B_j(y, r), \qquad y \in \Omega, 0 < r < \delta \, dist(y, \partial \Omega), 1 \leq j \leq n.$$

PROOF. We prove this using the same argument used in the proof of Proposition 83. The only real difference is that the flag balls fail to be monotone, satisfying instead the weak monotonicity condition

$$B(y, s) \subset B(y, r), \qquad \text{for } 0 < s < cr < r < \delta \, dist(y, \partial \Omega).$$

However, the reverse doubling property (215) of the side lengths $B_j(x, r)$ of the rectangles $B(x, r)$ will compensate for this lack of monotonicity.

Suppose that $y = 0$, $0 < r < \delta \, dist(0, \partial\Omega)$ and set $B_j(r) = B_j(y, r)$, $B(r) = B(y, r)$ for $1 \leq j \leq n$. For $\frac{1}{2} r < t \leq r$, let $R_j(t) = \frac{t}{r} B_j(r)$ and consider the rectangles

$$R(t) = (-t, t) \times \prod_{j=2}^{n} (-R_j(t), R_j(t)),$$

which satisfy $B(c_0 r) \subset R(t) \subset B(r)$. Note that $R(t)$ is the rectangle $B(r)$ dilated by the factor $\frac{t}{r}$. Given $\frac{1}{2} r < s < t \leq r$, define

$$\psi(x) = \prod_{j=1}^{n} \varphi\left(\frac{R_j(t) - |x_j|}{R_j(t) - R_j(s)} \right),$$

where

$$\varphi(\xi) = \begin{cases} 0, & \xi \leq 0 \\ \xi, & 0 \leq \xi \leq 1 \\ 1, & 1 \leq \xi \end{cases}.$$

We compute by the chain rule and the support conditions on φ that

$$|\nabla_a \psi(x)| \leq \sum_{j=1}^{n} a_j(x) \chi_{\{x \in B(r) : R_j(s) \leq |x_j| \leq R_j(t)\}} \frac{1}{R_j(t) - R_j(s)},$$

Since a_j is reverse Hölder of order p in the first variable x_1, we have for $j \geq 2$,

$$\left(\frac{1}{2r} \int_{y_1 - r}^{y_1 + r} a_j(x_1, x')^p \, dx_1 \right)^{\frac{1}{p}} \leq \frac{C}{r} \int_{y_1 - r}^{y_1 + r} a_j(x_1, x') \, dx_1 \leq \frac{C}{r} A_j((y_1, x'), r) \leq \frac{C}{r} B_j(r),$$

where the final inequality follows from the definition of $B_j(r)$. Just as in the proof of Proposition 83, Minkowski's inequality now yields

$$\left(\frac{1}{|B(r)|}\int|\nabla_a\psi|^p\right)^{\frac{1}{p}} \le \frac{C}{r}\sum_{j=1}^{n}\left\{\frac{B_j(r)}{\frac{t}{r}B_j(r)-\frac{s}{r}B_j(r)}\right\}^{\frac{1}{p'}}$$

$$= \frac{Cn}{r}\left(\frac{r}{t-s}\right)^{\frac{1}{p'}},$$

Now let $r_j - r_{j+1} = \frac{r}{5(j+1)^2}$, $r_0 = r$, $s = r_{j+1}$, $t = r_j$ and write ψ_j for ψ to obtain

$$\left(\frac{1}{|B(r)|}\int|\nabla_a\psi_j|^p\right)^{\frac{1}{p}} \le \frac{Cn}{r}\left(\frac{r_0}{r_j-r_{j+1}}\right)^{\frac{1}{p'}} = \frac{Cn}{r}(j+1)^{\frac{2}{p'}}.$$

It is now easy to verify that the sequence $\{\psi_{2j}\}_{j=1}^{\infty}$ satisfies the conditions in the "accumulating sequence of Lipschitz cutoff functions" condition (20) with $N = \frac{2}{p'}$. This completes the proof of Proposition 84.

The remainder of the reduction of the proof of Theorem 23 is now carried out in the same way as that of Theorem 24.

In this section, we have reduced the proofs of Theorems 17 and 23 (respectively 20 and 24) to proving that the flag balls $\mathcal{B}$ (respectively the noninterference balls $\mathcal{A}$) satisfy the reverse doubling condition (215) (respectively the size-limiting condition (91)) and form a δ-local prehomogeneous (respectively δ-local homogeneous) space adapted to $\mathcal{X}$ (note that our proof of Theorem 17 in subsection 4.5 required condition (91) for the flag balls, but our proof of the stronger Theorem 23 in subsection 4.5.1 did not - nonetheless, we will establish (91) for the flag balls in the next section). These latter properties will be proved in the next section.

CHAPTER 5

Homogeneous spaces and subrepresentation inequalities

The purpose of this section is to show that for some $\delta > 0$, the family of noninterference balls $\{A(x,r)\}_{x \in \Omega, 0 < r < \delta \, dist(x, \partial\Omega)}$ is a δ-local homogeneous space satisfying the size-limiting condition (91) that is adapted to $\mathcal{X}$, and also that the family of flag balls $\{B(x,r)\}_{x \in \Omega, 0 < r < \delta \, dist(x, \partial\Omega)}$ is a δ-local prehomogeneous space satisfying the reverse doubling condition (215) that is adapted to $\mathcal{X}$. The main restrictions on the size of $\delta > 0$ will arise in making sense of the definitions of noninterference and flag balls, as well as in making sense of the statements of the various lemmas below (where the supremum of quantities such as $A_j(x,r)$ are taken over balls $B(x, Cr)$ for $0 < Cr < \delta \, dist(x, \partial\Omega)$).

1. The noninterference balls

Let $a(x) = (1, a_2(x), \dots, a_n(x))$, $x = (x_1, \dots, x_n) \in \Omega$, be as above, except that we do *not* assume the continuity of $a_j(x)$ in x. In particular, there is a doubling constant C_d so that if I and J are appropriate one-dimensional intervals with $J \subset 5I$, then

$$\int_J a_i(x_1, x_2, \dots, x_n) \, dx_1 \le C_d \int_I a_i(x_1, x_2, \dots, x_n) \, dx_1$$

for each i uniformly in $x_2, \dots, x_n$, and there is a Lipschitz constant C_L so that

$$|a_i(x_1, x_2, \dots, x_n) - a_i(x_1, x_2', \dots, x_n')| \le C_L \max_{j=2,\dots,n} |x_j - x_j'|,$$

for each i uniformly in x_1. Recall that given $i = 2, \dots, n$, $x \in \Omega$ and $t \in (-\infty, \infty)$, $A_i(x,t)$ is defined by

$$A_i(x,t) = \int_0^t a_i(x_1 + s, x_2, \dots, x_n) \, ds.$$

Note that $A_i(x,t)$ has the same sign as t. This definition makes sense provided $|t| < dist(x, \partial\Omega)$.

We assume in this section that $A_i(x,t) \ne 0$ if $t \ne 0$, for all x and i. Due to doubling of a_i in the first variable, this is equivalent to assuming that $A_i(x,r) > 0$ for all x when $r > 0$. The n-dimensional rectangles $A(x,r)$, $r > 0$, are defined by

$$(216) \qquad A(x,r) = (x_1 - r, x_1 + r) \times \prod_{i=2}^n (x_i - A_i(x,r), x_i + A_i(x,r)),$$

and are contained in Ω provided $0 < r < \delta \, dist(x, \partial\Omega)$ for $\delta > 0$ sufficiently small depending on $C_{\max}$ in (37). Note that these rectangles are nonempty, open and bounded.

In addition to the doubling and Lipschitz conditions just mentioned, we also assume in this subsection that each $A_i(x, r)$ satisfies the following noninterference condition (Definition 18 in the introduction): there are positive constants C_c and δ such that for all $x \in \Omega$ and all r with $0 < r < \delta \, dist(x, \partial\Omega)$,

$$(217) \qquad C_c^{-1} A_i(x, r) \leq A_i(z, r) \leq C_c A_i(x, r), \qquad z \in A(x, r).$$

Note that $A_i(z, r)$ is defined if $\delta > 0$ is sufficiently small depending on $C_{\max}$ in (37). If all the a_i are the same, including the case $n = 2$ (when there is only one a_i), it turns out that (217) is automatically true - see Lemma 111 in the appendix; moreover, as we will show in Lemma 109 in the appendix, (217) is a corollary of the *strong* noninterference condition (41). Condition (217) implies that if $0 < r < \delta \, dist(x, \partial\Omega)$, then

$$C_c^{-(n-1)} |A(x, r)| \leq |A(z, r)| \leq C_c^{n-1} |A(x, r)|, \qquad z \in A(x, r),$$

for the same constant C_c as in (217).

As x and r vary, the rectangles $A(x, r)$ have an engulfing property, which is a precursor to (56), that is described in the next lemma. If c is a positive constant, $cA(x, r)$ will denote the rectangle with center x whose dimensions are c times the corresponding dimensions of $A(x, r)$, i.e., $cA(x, r)$ is the Euclidean dilation of $A(x, r)$ by the factor c in each coordinate direction, as opposed to the rectangle $A(x, cr)$.

We remind the reader of the convention regarding $\delta > 0$ in subsection 1.1 of the introduction.

LEMMA 85. *Suppose that each a_i is doubling in x_1 uniformly in $x_2, \ldots, x_n$ and Lipschitz continuous in $x_2, \ldots, x_n$ uniformly in x_1, and that (217) holds for some positive δ. If $A(x, r)$ and $A(y, s)$ are rectangles of type (216) with $A(x, r) \cap A(y, s) \neq \emptyset$ and $0 < s \leq r < \delta \, dist(x, \partial\Omega)$, $s < \delta \, dist(y, \partial\Omega)$, then $A(y, s) \subset cA(x, r)$ with $c = 1 + 2C_c^2$.*

PROOF. Since $A(y, s)$ and $A(x, r)$ intersect, $A(y, s)$ is contained in the rectangle with center x whose half-dimensions are $r + 2s$ and $A_i(x, r) + 2A_i(y, s)$, $i = 2, \ldots, n$. If z is a point in $A(y, s) \cap A(x, r)$, then by (217), $A_i(y, s) \leq C_c A_i(z, s)$ and $A_i(z, r) \leq C_c A_i(x, r)$ for all i. Also, $A_i(z, s) \leq A_i(z, r)$ since $s \leq r$. Thus,

$$A_i(y, s) \leq C_c A_i(z, s) \leq C_c A_i(z, r) \leq C_c^2 A_i(x, r).$$

Therefore, $A_i(x, r) + 2A_i(y, s) \leq (1 + 2C_c^2) A_i(x, r)$, and since also $r + 2s \leq 3r \leq (1 + 2C_c^2)r$, the lemma follows.

It is simple to show that since each $a_i(x)$ satisfies the doubling condition with constant C_d in the first variable uniformly in the others, then there is a positive constant d, depending only on C_d, such that if I and J are appropriate intervals in $(-\infty, \infty)$ with $I \subset J$,

$$(218) \qquad \int_J a_i(x_1, x_2, \ldots, x_n) \, dx_1 \leq C_d \left(\frac{|J|}{|I|} \right)^d \int_I a_i(x_1, x_2, \ldots, x_n) \, dx_1$$

for all i uniformly in $x_2, \ldots, x_n$. In fact we can choose $d = \ln_5 C_d$. Moreover, as is well-known, it follows that each $a_i(x)$ satisfies a *reverse* doubling condition in the first variable uniformly in the others, i.e., there are positive constants $\tilde{d}, \widetilde{C_d}$, again depending only on C_d, such that if I and J are intervals with $I \subset J$, then

$$(219) \qquad \int_J a_i(x_1, x_2, \ldots, x_n)\, dx_1 \geq \widetilde{C_d} \left(\frac{|J|}{|I|} \right)^{\tilde{d}} \int_I a_i(x_1, x_2, \ldots, x_n)\, dx_1$$

for all i uniformly in $x_2, \ldots, x_n$.

It follows immediately from the definition of $A_i(x, \cdot)$ and from (218) and (219) that $A_i(x, \cdot)$ satisfies the following doubling and reverse doubling estimates for each i:

$$(220)$$

$$\widetilde{C_d} \left(\frac{r}{s} \right)^{\tilde{d}} A_i(x, s) \leq A_i(x, r) \leq C_d \left(\frac{r}{s} \right)^{d} A_i(x, s) \quad \text{if } 0 < s \leq r < \delta\, dist(x, \partial\Omega)$$

uniformly in x. Note that (220) implies that $\widetilde{C_d} \leq 1 \leq C_d$ and that $\tilde{d} \leq d$. We always assume throughout this section that $A_i(x, r) > 0$ if $r > 0$, but (220) does not require this assumption; in fact, (220) implies that if $A_i(x, r) = 0$ for some $r > 0$, then $A_i(x, r) = 0$ for all r, so that a_i vanishes identically on the line parallel to the x_1-axis through x.

1.1. Adapted noninterference balls.

Combining Lemma 85 with the reverse doubling inequality in (220) yields the δ-local engulfing property (68) for the rectangles $A(x, r)$. Indeed, if $A(x, r) \cap A(y, r) \neq \phi$, then $A(y, r) \subset cA(x, r)$ by Lemma 85. By the first inequality in (220), $cA(x, r) \subset A(x, \gamma r)$ if we choose γ so that $c = \widetilde{C_d}\gamma^{\tilde{d}}$. The doubling inequality in (220) yields the δ-local doubling property (71) with $C_{doub} = 2^{dn} C_d^n$. The δ-local monotonicity and scale properties (72) and (69) are immediate here, and as we have already observed, the rectangles $A(x, r)$ are nonempty and open. This completes the demonstration that the family of noninterference balls $\mathcal{A} = \{A(x, r)\}_{x \in \mathbb{R}^n, 0 < r < \delta\, dist(x, \partial\Omega)}$ forms a δ-local homogeneous space $\mathbb{A} = (\Omega, \mathcal{A})$ on Ω.

The noninterference condition in Definition 18 implies both the size-limiting condition (91) and that the vector fields $X_j = a_j(x)\frac{\partial}{\partial x_j}$, $1 \leq j \leq n$, are adapted to the noninterference balls in the sense of Definition 78. The proofs of Theorems 20 and 24 are now complete as mentioned at the end of section 4.

REMARK 86. *Under the hypotheses of Theorem 20 or 24, the ratio $\frac{d(x, x')}{\rho}$ in the Hölder estimate (175) of Theorem 65 can be replaced with the larger quantity*

$$\max \left\{ \frac{|x_1 - x_1'|}{\rho}, \left(\frac{|x_2 - x_2'|}{A_2(y, \rho)} \right)^{\frac{1}{d}}, \ldots, \left(\frac{|x_n - x_n'|}{A_n(y, \rho)} \right)^{\frac{1}{d}} \right\},$$

where d is a doubling exponent as in (218). Indeed, the quasimetric $d(x, x')$ in (175) is the symmetric quasimetric appearing in Theorem 8, which in turn arises from an application of Lemma 37 to the quasimetric associated to the extension in Proposition 41 of the δ-local homogeneous space $\mathbb{A}$. The above estimate now follows from a calculation using the definition (59) of quasimetric, the monotonicity (72) of the noninterference balls in $\mathbb{A}$, and the second inequality in (220).

2. The flag balls

We now turn to showing that the flag balls satisfy the reverse doubling condition (215), and form a δ-local prehomogeneous space that is adapted to the vector fields $X_j = a_j(x)\frac{\partial}{\partial x_j}$ in the sense of Definition 78 under the hypotheses of either Theorem

17 or 23. As mentioned at the end of section 4, this will complete the proofs of Theorems 17 and 23. We now assume $n \geq 3$ since otherwise the flag balls coincide with the noninterference balls which have already been treated. The constants C_d and C_L are defined at the beginning of subsection 5.1, and the term "geometric constant" means a constant which only depends on n, C_d and C_L. We consider vector fields $\frac{\partial}{\partial x_1}, a_2(x)\frac{\partial}{\partial x_2}, ..., a_n(x)\frac{\partial}{\partial x_n}$ for $x \in \Omega$, where the a_j are nonnegative functions doubling in the first variable x_1 uniformly in $x_2, ..., x_n$, and Lipschitz continuous in $x_2, ..., x_n$ uniformly in x_1. We emphasize again that we do *not* assume the continuity of the $a_j(x)$ in x. We will however replace the flag condition with a more basic nondegeneracy assumption below. Our goal in this subsection is to show that in the presence of this nondegeneracy condition, the flag balls $B(x,r)$ satisfy the δ-local prehomogeneous space properties of engulfing (68), weak monotonicity (70), scale (69) and doubling (71), to also show that the rectangles $\widetilde{B}(x,r)$ satisfy the doubling condition (180) and the size-limiting condition (91), as well as the reverse doubling condition (215) and the reverse size condition 2 in Definition 78. To accomplish all of this, we will need to impose additional conditions on the functions a_j as we go. However, we no longer assume in this section that $A_j(x,t) > 0$ for $t > 0$, or that (217) holds.

We now redefine the flag balls introduced in the introduction, but in more detail here. We begin by defining the $B_j(x,r)$, $j = 2, ..., n$. Fix x and r with $x \in \Omega$, $0 < r < \delta \, dist(x, \partial\Omega)$, and define a rearrangement $\{j_2, ..., j_n\}$ of $\{2, ..., n\}$ and corresponding functions $\{B_{j_k}(x,r)\}_{k=2}^n$ inductively as follows. First, pick the least $j_2 \in \{2, ..., n\}$ with $A_{j_2}(x,r) \geq A_j(x,r)$ for all $j = 2, ..., n$, and let

$$(221) \qquad B_{j_2}(x,r) = A_{j_2}(x,r) = \max_j A_j(x,r).$$

For the next stage, we use $B_{j_2}(x,r)$ to fill-out the remaining $A_j(x,r)$ in the variable x_{j_2}, i.e., for each $j \neq j_2$, we consider the quantity

$$\max_{z_i = x_i \text{ if } i \neq j_2; \ |z_{j_2} - x_{j_2}| \leq B_{j_2}(x,r)} A_j(z,r),$$

and then pick the least $j_3 \in \{2, ..., n\}$, $j_3 \neq j_2$, so that

$$\max_{z_i = x_i \text{ if } i \neq j_2; \ |z_{j_2} - x_{j_2}| \leq B_{j_2}(x,r)} A_{j_3}(z,r) \geq \max_{z_i = x_i \text{ if } i \neq j_2; \ |z_{j_2} - x_{j_2}| \leq B_{j_2}(x,r)} A_j(z,r)$$

for all $j \neq j_2$. Define

$$B_{j_3}(x,r) = \max_{z_i = x_i \text{ if } i \neq j_2; \ |z_{j_2} - x_{j_2}| \leq B_{j_2}(x,r)} A_{j_3}(z,r)$$

$$= \max_{j \neq j_2} \left(\max_{z_i = x_i \text{ if } i \neq j_2; \ |z_{j_2} - x_{j_2}| \leq B_{j_2}(x,r)} A_j(z,r) \right).$$

Having defined $B_{j_2}(x,r), ..., B_{j_{k-1}}(x,r)$, pick the least $j_k \neq j_2, ..., j_{k-1}$ with

$$\max_{z_i = x_i, i \notin \{j_2, ..., j_{k-1}\}; \ |z_i - x_i| \leq B_i(x,r), i \in \{j_2 ..., j_{k-1}\}} A_{j_k}(z,r)$$

equal to or greater than

$$\max_{z_i = x_i, i \notin \{j_2, ..., j_{k-1}\}; \ |z_i - x_i| \leq B_i(x,r), i \in \{j_2, ..., j_{k-1}\}} A_j(z,r)$$

for all $j \neq j_2, \ldots, j_{k-1}$, and define

(222) $B_{j_k}(x, r)$

$$= \max_{z_i = x_i, i \notin \{j_2, \ldots, j_{k-1}\}; \, |z_i - x_i| \leq B_i(x,r), i \in \{j_2, \ldots, j_{k-1}\}} A_{j_k}(z, r)$$

$$= \max_{j \notin \{j_2, \ldots, j_{k-1}\}} \left(\max_{z_i = x_i, i \notin \{j_2, \ldots, j_{k-1}\}; \, |z_i - x_i| \leq B_i(x,r), i \in \{j_2, \ldots, j_{k-1}\}} A_j(z, r) \right).$$

All of these definitions make sense provided $x \in \Omega$ and $0 < r < \delta \, dist(x, \partial\Omega)$ for $\delta > 0$ sufficiently small depending on $C_{\max}$ in (37). For convenience in notation, we will refrain from writing " $x \in \Omega$ and $0 < r < \delta \, dist(x, \partial\Omega)$" for most of the remainder of the paper, but such a restriction will always be in force.

REMARK 87. *Note that if we assume the $a_j(x)$ are continuous in x (or even just in x_1 if $a_1 \equiv 1$), then the flag condition in Definition 13 holds if and only if each $B_j(x, r) > 0$ when $r > 0$. Consequently, even when the $a_j(x)$ are not continuous, we will always assume throughout this subsection the nondegeneracy condition that each $B_j(x, r) > 0$ when $r > 0$.*

Note that (222) holds only when $k \geq 3$ and makes no sense when $k = 2$ because of the pattern of the definition. When $k \geq 3$, it implies that if $j = j_k, j_{k+1}, \ldots, j_n$ and if z is a point whose components satisfy $z_i = x_i$ for $i = 1, j_k, \ldots, j_n$ and $|z_i - x_i| \leq B_i(x, r)$ for $i = j_2, \ldots, j_{k-1}$, then

$$B_{j_k}(x, r) \geq A_j(z, r).$$

In the remaining case $k = 2$, the last estimate of course holds with $z = x$, i.e.,

$$B_{j_2}(x, r) \geq A_j(x, r) \quad \text{for all } j \geq 2.$$

The order of the permutation $\{j_2, \ldots, j_n\}$ generally depends on x and r, but we have not attempted to show this in our notation. The simplest notation arises in the special case when $j_2 = 2, \ldots, j_n = n$, i.e., when there is no permutation in the natural order. We then have

(223) $B_2(x, r) = A_2(x, r),$

$$B_j(x, r) = \max_{|z_i - x_i| \leq B_i(x,r), i = 2, \ldots, j-1} A_j((x_1, z_2, \ldots, z_{j-1}, x_j, \ldots, x_n), r).$$

In this special case, we also have

(224) $\begin{aligned} B_j(x, r) &\geq A_k((x_1, z_2, \ldots, z_{j-1}, x_j, \ldots, x_n), r) \\ for \quad k &\geq j \geq 3, \, |z_i - x_i| \leq B_i(x, r), \\ B_2(x, r) &\geq A_k(x, r) \quad for \, k \geq 2. \end{aligned}$

For simplicity of notation, we often assume in the proofs below that the ordering is the natural one, unless of course more than one pair x, r is being considered. In case $x = 0$, we usually write $B_{j_k}(r)$ instead of $B_{j_k}(0, r)$. The next lemma shows that the sequence $\{B_{j_k}(x, r)\}_{k=2}^n$ is essentially monotone decreasing in k.

LEMMA 88. *Let $\{B_j(x, r)\}$ be defined as above in the order $B_{j_2}(x, r), \ldots, B_{j_n}(x, r)$. Then*

$$B_{j_{k+1}}(x, r) \leq (1 + rC_L) B_{j_k}(x, r), \quad k = 2, \ldots, n - 1.$$

Consequently, if $i > k$ then $B_{j_i}(x, r) \leq (1 + rC_L)^{n-2} B_{j_k}(x, r)$.

PROOF. The second statement follows easily from the first. To prove the first one, consider the case of the natural ordering and $x = 0$. Fixing j and denoting $B_j(0, r) = B_j(r)$, we have from (223) that there exist $z_2, \ldots, z_j$ with $|z_i| \leq B_i(r)$ and

$$B_{j+1}(r) = A_{j+1}((0, z_2, \ldots, z_j, 0, \ldots, 0), r).$$

In case $j = 2$, we obtain

$$
\begin{aligned}
B_3(r) &\leq |A_3((0, z_2, 0, \ldots, 0), r) - A_3(0, r)| + A_3(0, r) \\
&\leq rC_L|z_2| + B_2(r) \leq (1 + rC_L)B_2(r),
\end{aligned}
$$

where we have used the second part of (224). If $j \geq 3$,

$$
\begin{aligned}
B_{j+1}(r) &\leq |A_{j+1}((0, z_2, \ldots, z_j, 0, \ldots, 0), r) - A_{j+1}((0, z_2, \ldots, z_{j-1}, 0, \ldots, 0), r)| \\
&\quad + A_{j+1}((0, z_2, \ldots, z_{j-1}, 0, \ldots, 0), r) \\
&\leq rC_L|z_j| + A_{j+1}((0, z_2, \ldots, z_{j-1}, 0, \ldots, 0), r) \\
&\leq rC_L B_j(r) + \max_{|\xi_i| \leq B_i(r)} A_{j+1}((0, \xi_2, \ldots, \xi_{j-1}, 0, \ldots, 0), r) \\
&\leq rC_L B_j(r) + B_j(r) = (1 + rC_L)B_j(r),
\end{aligned}
$$

by (224) which completes the proof.

The functions $B_j(x, r)$ are defined in a chosen order with some (but not all) of the variables x_i filled-out. The next lemma shows that the remaining variables can also be filled-out in a similar way without much effect.

LEMMA 89. *If $C \geq 1$, then*

$$B_j(x, r) \approx \max_{z:|z_1 - x_1| \leq Cr;\ |z_i - x_i| \leq B_i(x, r), i \geq 2} A_j(z, r).$$

The constants of equivalence are independent of x and r; they are geometric constants which also depend on C.

PROOF. The fact that the left side of the conclusion is at most the right side is obvious from the definition of $B_j(x, r)$. To show the opposite inequality (with an appropriate constant factor), suppose we have the natural order and $x = 0$. Then by definition,

$$
\begin{aligned}
B_j(r) &= \max_{|\xi_i| \leq B_i(r)} A_j((0, \xi_2, \ldots, \xi_{j-1}, 0, \ldots, 0), r), \qquad j \geq 3, \\
B_2(r) &= A_2(0, r).
\end{aligned}
$$

Fix $C \geq 1$ and let $z = (z_1, \ldots, z_n)$ with $|z_1| \leq Cr$ and $|z_i| \leq B_i(r)$ for $i \geq 2$. By doubling in the first variable,

$$A_j(z, r) \leq cA_j((0, z_2, \ldots, z_n), r).$$

Thus if $j \geq 3$, since $|z_i| \leq B_i(r)$ for $i \geq 2$,

$$
\begin{aligned}
A_j(z,r) &\leq c|A_j((0,z_2,\ldots,z_n),r) - A_j((0,z_2,\ldots,z_{j-1},0\ldots,0),r)| \\
&\quad + cA_j((0,z_2,\ldots,z_{j-1},0,\ldots,0),r) \\
&\leq crC_L \sum_{i=j}^{n} |z_i| + cA_j((0,z_2,\ldots,z_{j-1},0,\ldots,0),r) \\
&\leq crC_L \sum_{i=j}^{n} B_i(r) + cB_j(r) \\
&\leq crC_L(n-1)(1+rC_L)^{n-2}B_j(r) + cB_j(r) \\
&\leq c'B_j(r),
\end{aligned}
$$

by Lemma 88. This completes the proof when $j \geq 3$.

In the remaining case $j = 2$, we will show the better estimate

$$
\max_{|z_1| \leq Cr;\ |z_i| \leq CB_2(r), i \geq 2} A_2(z,r) \leq cB_2(r).
$$

In fact, if $|z_1| \leq Cr$ and $|z_i| \leq CB_2(r)$ for $i \geq 2$, then

$$
\begin{aligned}
A_2(z,r) &\leq cA_2((0,z_2,\ldots,z_n),r) \\
&\leq c|A_2((0,z_2,\ldots,z_n),r) - A_2(0,r)| + cA_2(0,r) \\
&\leq crC_L \sum_{2}^{n} |z_i| + cB_2(r) \leq crC_L(n-1)CB_2(r) + cB_2(r) \leq c'B_2(r),
\end{aligned}
$$

which completes the proof.

In Lemma 89, the first variable was filled-out by an amount Cr with $C \geq 1$, but the other variables were filled-out by only $B_i(r)$, as opposed to $CB_i(r), C > 1$. We now assume an extra condition which will allow us to fill-out all the variables by larger amounts, and which will also be useful later for proving the doubling property of $B_j(x,r)$ in r for all x. In fact, we will assume that given $C \geq 1$, there is a constant c so that for all j, x and r,

$$
(225) \quad \max_{|z_1-x_1| \leq Cr;\ |z_i-x_i| \leq CB_i(x,r), i \geq 2} A_j(z,r)
$$
$$
\leq c\frac{1}{r\prod_{i=2}^{n} B_i(x,r)} \int_{|z_1-x_1| \leq r;\ |z_i-x_i| \leq B_i(x,r), i \geq 2} A_j(z,r)\, dz.
$$

Note that since $C \geq 1$, the inequality opposite to (225) is obvious with $c = 1$. In subsection 6.5 of the appendix, we show that (225) holds if *either* of the following two conditions is satisfied:

either each a_j satisfies the RH_∞ condition in each variable $x_i, i \neq j, 1$, uniformly in the other variables, i.e., there is a constant C such that for $i,j = 2,\ldots,n, i \neq j$, and all one-dimensional intervals I which contain x_i,
$$
(226)
$$
$$
a_j(x_1,\ldots,x_n) \leq C\frac{1}{|I|}\int_I a_j(x_1,\ldots,x_{i-1},z_i,x_{i+1},\ldots,x_n)\, dz_i, \quad i,j \geq 2, i \neq j,
$$

uniformly in x_k for $k \neq i$,

or the $B_j(x,r)$ satisfy a strong non-interference condition of the form

$$(227) \qquad r \left(\sup_{z:|z_1-x_1|\leq Cr;\, |z_k-x_k|\leq CB_k(x,r),k\geq 2} \left| \frac{\partial a_j}{\partial z_i}(z) \right| \right) B_i(x,r)$$

$$\leq \widetilde{C_0} B_j(x,r), \quad i,j \geq 2, i \neq j$$

for all x and r and a suitably large geometric constant C. See Remark 19 for the interpretation of sup in (227). Note that we never consider the partial derivative $\partial/\partial z_1$ in (227). Recall that all $B_j(x,r) > 0$ if $r > 0$ by assumption. In proving Theorem 17, we will only use that (226) implies (225). In subsection 6.5 of the appendix, we describe in Theorem 108 a variant of Theorem 17 that uses the implication (227) $\Longrightarrow$ (225).

Assuming (225), we now easily obtain the following result.

LEMMA 90. *If (225) holds, then for $C \geq 1$ and $j = 2, \ldots, n$,*

$$B_j(x,r) \approx \max_{z:|z_1-x_1|\leq Cr;\, |z_i-x_i|\leq CB_i(x,r),i\geq 2} A_j(z,r).$$

The constants of equivalence are independent of x and r; they are geometric constants which also depend on C and the constants in (225).

PROOF. Let $C \geq 1$. The left side of (225) is at least $B_j(x,r)$ by definition of $B_j(x,r)$, and the right side of (225) is at most

$$\max_{|z_1-x_1|\leq r;\, |z_i-x_i|\leq B_i(x,r),i\geq 2} A_j(z,r),$$

which is bounded by $cB_j(x,r)$ by Lemma 89. The conclusion then follows immediately from (225).

Thus, (225) implies that $B_j(x,r)$ is equivalent to either side of (225), and in particular that
(228)

$$B_j(x,r) \approx \frac{1}{r\prod_2^n B_i(x,r)} \int_{|z_1-x_1|\leq r;\, |z_i-x_i|\leq B_i(x,r)} A_j(z,r)\, dz_1 \ldots dz_n, \quad j \geq 3.$$

In fact, the domain of integration on the right can be enlarged by constant factors. In the special case $j = j_2$, we have a better result *without assuming (225)*: given a constant C, there is a geometric constant r_1 which also depends on C so that if $0 < r < r_1$, then
(229)

$$B_{j_2}(x,r) \approx A_{j_2}(z,r) \quad \text{if } z \text{ satisfies } |z_1-x_1| \leq Cr, |z_j-x_j| \leq CB_{j_2}(x,r), j \geq 2.$$

Recall that $B_j(x,r) \leq cB_{j_2}(x,r)$ for all j by Lemma 88. The proof of (229) is similar to part of the proof of Lemma 89. In fact, we showed there that if $x = 0$ and $j_2 = 2$, then

$$\max_{|z_1|\leq Cr;\, |z_i|\leq CB_2(r),i\geq 2} A_2(z,r) \leq cB_2(r),$$

while if z satisfies $|z_1| \leq Cr$ and $|z_i| \leq CB_2(r)$ for $i \geq 2$, then we also have

$$
\begin{aligned}
B_2(r) &= A_2(0,r) \leq cA_2((z_1,0,\ldots,0),r) \\
&\leq c|A_2((z_1,0,\ldots,0),r) - A_2((z_1,z_2,\ldots,z_n),r)| + cA_2((z_1,\ldots,z_n),r) \\
&\leq cC_Lr\sum_2^n |z_i| + cA_2(z,r) \\
&\leq cC_Lr\sum_2^n CB_2(r) + cA_2(z,r) \\
&\leq cC_L(n-1)CrB_2(r) + cA_2(z,r).
\end{aligned}
$$

Choosing r small and absorbing the term $cC_L(n-1)CrB_2(r)$, we obtain $B_2(r) \leq c'A_2(z,r)$, and therefore,

$$
B_2(r) \leq c' \min_{|z_1|\leq Cr;\ |z_j|\leq CB_2(r),j\geq 2} A_2(z,r),
$$

and (229) follows.

REMARK 91. *If (228) holds (e.g., if (225) holds), there is a variant of (228) which involves integration over a smaller set. To state it, fix x and r and suppose that the $B_j(x,r)$ are defined in the order $\{j_k : k = 2,\ldots,n\}$. Then if $k \geq 2$ and r is small,*

$$
\begin{aligned}
B_{j_k}(x,r) &\approx \frac{1}{\prod_{j=j_2,\ldots,j_{k-1}} B_j(x,r)} \\
&\times \int_{|z_j-x_j|\leq B_j(x,r)} A_{j_k}(z,r)\, dz_{j_2}\ldots dz_{j_{k-1}}\, |_{z_i=x_i,i=1,j_k,\ldots,j_n}.
\end{aligned}
$$

Moreover, in this integral average, the fixed variables $z_1, z_{j_k},\ldots,z_{j_n}$ that are chosen above to be $x_1, x_{j_k},\ldots,x_{j_n}$, respectively, can instead be chosen to be any values $y_1, y_{j_k},\ldots,y_{j_n}$ with $|y_1 - x_1| \leq r$ and $|y_j - x_j| \leq B_j(x,r), j = j_k,\ldots,j_n$, provided r is small. To see this, suppose $x = 0$ and that we have the natural order. Fix $j \geq 2$ and let $|y_1| \leq r, |y_j| \leq B_j(r),\ldots,|y_n| \leq B_n(r)$. First note that by doubling in the first variable, we may drop the part of the average in z_1 and replace z_1 by y_1, i.e., by (228) and doubling,

$$
B_j(r) \approx \frac{1}{\prod_{i=2}^n B_i(r)} \int_{|z_i|\leq B_i(r)} A_j((y_1,z_2,\ldots,z_n),r)\, dz_2\ldots dz_n.
$$

Next, the absolute value of

$$
\frac{1}{\prod_{i=2}^n B_i(r)} \int_{|z_i|\leq B_i(r)} A_j((y_1,z_2,\ldots,z_n),r)\, dz_2\ldots dz_n -
$$

$$
\frac{1}{\prod_{i=2}^{j-1} B_i(r)} \int_{|z_i|\leq B_i(r);\ i=2,\ldots,j-1} A_j((y_1,z_2,\ldots,z_{j-1},y_j,\ldots,y_n),r)\, dz_2\ldots dz_{j-1}
$$

is at most

$$
\sum_j^n C_Lr|z_i - y_i| \leq \sum_j^n C_Lr2B_i(r) \leq crB_j(r)
$$

by Lemma 88. For small r, the desired estimate now follows by combining estimates and absorbing the term $crB_j(r)$. In case $k = 2$, (229) gives an even stronger equivalence without assuming (228) or (225).

2.1. Doubling and monotonicity properties of flag balls. The next lemma establishes doubling and reverse doubling properties of $B_j(x,r)$ like those given in (220) for $A_j(x,r)$. In the lemma, we will assume that each $a_j(x)$ satisfies the doubling condition in each variable $x_i, i \geq 1, i \neq j$ (uniformly in the other variables), i.e., there is a constant C such that for all one-dimensional intervals I,

$$\int_{2I} a_j(x_1, \ldots, x_n)\, dx_i \leq C \int_I a_j(x_1, \ldots, x_n)\, dx_i \quad j \geq 2, i \geq 1, i \neq j,$$

uniformly in $x_k, k \neq i$. We leave it to the reader to verify that this is equivalent to assuming that each $A_j(x,r)$ satisfies the doubling condition in each variable $x_i, i \geq 1, i \neq j$, uniformly in r and the other variables. In fact, the corresponding constants C can be chosen to be equal.

LEMMA 92. *If (225) holds and each $a_j(x)$ satisfies the doubling condition in each variable $x_i, i \neq j$, and is Lipschitz continuous in $x_2, \ldots, x_n$ uniformly in x_1, there are positive constants $\widetilde{d}, d^*$ and C so that*

$$(230) \qquad C^{-1} \left(\frac{r}{s}\right)^{\widetilde{d}} B_j(x,s) \leq B_j(x,r) \leq C \left(\frac{r}{s}\right)^{d^*} B_j(x,s) \quad \text{if } 0 < s < r.$$

In particular, $B_j(x,r)$ satisfies the weak monotonicity property

$$(231) \qquad\qquad B_j(x,s) \leq B_j(x,r) \quad \text{if } 0 < s < cr,$$

with $c = C^{-\frac{1}{d}}$, and so the flag balls $B(x,r)$ satisfy the weak monotonicity condition (60). The constants $\widetilde{d}, d^*$ and C are geometric constants which also depend on the constants in (225) and on the doubling constants of the $a_j(x)$ in each $x_i, i \neq j$.

PROOF. To show the first estimate in (230), namely $B_j(x,s) \leq C(s/r)^{\widetilde{d}} B_j(x,r)$, suppose $x = 0$ and fix $0 < s < r$. Assume that the $B_j(s)$ are defined in the natural order $2, \ldots, n$ while the $B_j(r)$ are defined in the order $j_2, \ldots, j_n$. We proceed by induction on j. If $j = 2$, then

$$\begin{aligned} B_2(s) &= A_2(s) \leq C \left(\frac{s}{r}\right)^{\widetilde{d}} A_2(r) \quad \text{by (220)} \\ &\leq C \left(\frac{s}{r}\right)^{\widetilde{d}} B_2(r) \quad \text{by definition of } B_2(r). \end{aligned}$$

Here we note that (220) does not require the fact that the $A_i(x,t) \neq 0$. Now let $j \geq 3$ and assume (by induction) that $B_i(s) \leq C(s/r)^{\widetilde{d}} B_i(r)$ for $i = 2, \ldots, j-1$ and $\widetilde{d}$ as in (220). Then by definition of $B_j(s)$,

$$\begin{aligned} B_j(s) &= \max_{|z_i| \leq B_i(s)} A_j((0, z_2, \ldots, z_{j-1}, 0, \ldots, 0), s) \\ &\leq \max_{|z_i| \leq C\left(\frac{s}{r}\right)^{\widetilde{d}} B_i(r)} A_j((0, z_2, \ldots, z_{j-1}, 0, \ldots, 0), s) \\ &\leq \max_{|z_i| \leq C B_i(r)} A_j((0, z_2, \ldots, z_{j-1}, 0, \ldots, 0), s). \end{aligned}$$

Since A_j satisfies the reverse doubling condition (220), we can continue with

$$\begin{aligned} &\leq \max_{|z_i| \leq C B_i(r)} C \left(\frac{s}{r}\right)^{\widetilde{d}} A_j((0, z_2, \ldots, z_{j-1}, 0, \ldots, 0), r) \\ &\leq C \left(\frac{s}{r}\right)^{\widetilde{d}} \max_{|z_i| \leq C B_i(r)} A_j((0, z_2, \ldots, z_n), r) \leq C \left(\frac{s}{r}\right)^{\widetilde{d}} B_j(r), \end{aligned}$$

by Lemma 90. This proves the first part of (230). Note that the value of $\tilde{d}$ is the same as in (220).

To prove the rest of (230), namely $B_j(x,r) \leq C(r/s)^{d^*} B_j(x,s)$, $0 < s < r$, we again suppose that $x = 0$, but now assume that the $B_j(r)$ are defined in the natural order $2, \ldots, n$ while the $B_j(s)$ are defined in the order $j_2, \ldots, j_n$. For $j = 2$,

$$B_2(r) = A_2(r) \leq C\left(\frac{r}{s}\right)^d A_2(s) \quad \text{by (220)}$$

$$\leq C\left(\frac{r}{s}\right)^d B_2(s) \quad \text{by definition of } B_2(s).$$

For $j \geq 3$, assuming by induction that $B_i(r) \leq C(r/s)^{d_1} B_i(s)$ for $i = 2, \ldots, j-1$ and some $d_1 > 0$, we obtain from (228) that

$$B_j(r) \leq \frac{c}{\prod_2^{j-1} B_i(r)} \int_{|z_i| \leq B_i(r)} A_j((0, z_2, \ldots, z_{j-1}, 0, \ldots, 0), r) \, dz_2 \ldots dz_{j-1}$$

$$\leq \frac{c}{\prod_2^{j-1} B_i(r)} \int_{|z_i| \leq C\left(\frac{r}{s}\right)^{d_1} B_i(s)} A_j((0, z_2, \ldots, z_{j-1}, 0, \ldots, 0), r) \, dz_2 \ldots dz_{j-1}$$

$$\leq \frac{c}{\prod_2^{j-1} B_i(r)} \left(\left(\frac{r}{s}\right)^{d_1}\right)^{d_2} \int_{|z_i| \leq B_i(s)} A_j((0, z_2, \ldots, z_{j-1}, 0, \ldots, 0), r) \, dz_2 \ldots dz_{j-1}$$

for some $d_2 > 0$ by doubling of $A_j(z,r)$ in each variable $z_i, i = 2, \ldots, j-1$ (in particular, $i \neq j$). Now, again by (220),

$$A_j((0, z_2, \ldots, z_{j-1}, 0, \ldots, 0), r) \leq C\left(\frac{r}{s}\right)^d A_j((0, z_2, \ldots, z_{j-1}, 0, \ldots, 0), s).$$

Combining estimates gives

$$B_j(r) \leq \frac{c}{\prod_2^{j-1} B_i(r)} \left(\frac{r}{s}\right)^{d_1 d_2 + d} \max_{|z_i| \leq B_i(s)} A_j((0, z_2, \ldots, z_{j-1}, 0, \ldots, 0), s) \prod_2^{j-1} B_i(s)$$

$$\leq c\left(\frac{r}{s}\right)^{d_1 d_2 + d - \tilde{d}} \max_{|z_i| \leq B_i(s)} A_j((0, z_2, \ldots, z_{j-1}, 0, \ldots, 0), s)$$

since

$$B_i(s) \leq c\left(\frac{s}{r}\right)^{\tilde{d}} B_i(r)$$

by what has already been proved. By Lemma 89, the max above is bounded by a multiple of $B_j(s)$, and we obtain

$$B_j(r) \leq c\left(\frac{r}{s}\right)^{d_1 d_2 + d - \tilde{d}} B_j(s).$$

The second estimate in (230) now follows if d^* is chosen sufficiently large.

Finally, the weak monotonicity estimate (231) follows from the first part of (230) since if $0 < s < cr$, $0 < c \leq 1$, then

$$B_j(x,s) \leq C\left(\frac{cr}{r}\right)^{\tilde{d}} B_j(x,r) = B_j(x,r)$$

if $Cc^{\tilde{d}} = 1$. This completes the proof of the lemma.

We next derive an analogue of the noninterference condition for the side lengths $B_j(x,r)$ and the corresponding rectangles $B(x,r)$ defined by

$$(232) \qquad B(x,r) = \{y : |y_1 - x_1| < r, |y_j - x_j| < B_j(x,r), j \geq 2\}.$$

LEMMA 93. *Let (225) hold and suppose each $a_j(x)$ satisfies the doubling condition in each variable $x_i, i \neq j$, and is Lipschitz continuous in $x_2, ..., x_n$ uniformly in x_1. If c_1 is a positive constant, there is a positive constant r_1 so that if $0 < r < r_1$ and $y \in B(x, c_1 r)$, then*

$$B_j(y, r) \approx B_j(x, r), \quad j \geq 2.$$

The constants of equivalence and r_1 are geometric constants which also depend on c_1 and the constants in the other conditions.

PROOF. Let $x = 0$ and $y \in B(0, c_1 r)$. Then y satisfies $|y_1| \leq c_1 r$ and $|y_i| \leq B_i(c_1 r)$ if $i \geq 2$, and so also $|y_i| \leq c' B_i(r)$ if $i \geq 2$ by (230). We will show that $B_j(y, r) \leq C B_j(r)$ by induction on j, assuming that the $B_j(y, r)$ are defined in the natural order. For $j = 2$,

$$B_2(y, r) = A_2(y, r) \leq C B_2(r)$$

by definition of $B_2(y, r)$ and Lemma 90. Now assume that $j \geq 3$ and $B_i(y, r) \leq C B_i(r)$ for $i = 2, \ldots, j - 1$. We have

$$
\begin{aligned}
B_j(y, r) &= \max_{|z_i - y_i| \leq B_i(y, r)} A_j((y_1, z_2, \ldots, z_{j-1}, y_j, \ldots, y_n), r) \\
&\leq \max_{|z_i - y_i| \leq C B_i(r)} A_j((y_1, z_2, \ldots, z_{j-1}, y_j, \ldots, y_n), r) \\
&\leq \max_{|z_i| \leq (C+c') B_i(r)} A_j((y_1, z_2, \ldots, z_{j-1}, y_j, \ldots, y_n), r)
\end{aligned}
$$

since $|z_i| \leq |z_i - y_i| + |y_i|$ and $|y_i| \leq c' B_i(r), i \geq 2$. Thus,

$$B_j(y, r) \leq \max_{|z_1| \leq c_1 r; \, |z_i| \leq (C+c') B_i(r), i \geq 2} A_j(z, r) \leq c'' B_j(r)$$

by Lemma 90. This completes the proof that $B_j(y, r) \leq C B_j(r)$ for $j \geq 2$.

To prove the opposite inequality $B_j(r) \leq C B_j(y, r)$ (with $x = 0$ and $y \in B(0, c_1 r)$), we again proceed by induction, but now we assume that the $B_j(r)$ (rather than the $B_j(y, r)$) are defined in the natural order. For $j = 2$ and r small,

$$B_2(r) \leq c A_2(y, r) \leq c B_2(y, r)$$

by (229) and (224). For $j \geq 3$, by doubling in the first variable,

$$
\begin{aligned}
B_j(r) &= \max_{|z_i| \leq B_i(r)} A_j((0, z_2, \ldots, z_{j-1}, 0, \ldots, 0), r) \\
&\leq \max_{|z_i| \leq B_i(r)} C_d A_j((y_1, z_2, \ldots, z_{j-1}, 0, \ldots, 0), r) \\
&\leq \max_{|z_i - y_i| \leq (1+c') C B_i(y, r)} C_d A_j((y_1, z_2, \ldots, z_{j-1}, 0, \ldots, 0), r)
\end{aligned}
$$

since if $|z_i| \leq B_i(r)$, then

$$|z_i - y_i| \leq |z_i| + |y_i| \leq (1 + c') B_i(r) \leq (1 + c') C B_i(y, r), \quad i = 2, \ldots, j - 1,$$

by the inductive hypothesis. By applying Lemma 90 to the last max, we obtain $B_j(r) \leq c'' B_j(y, r)$, and the lemma is proved.

The rectangles $B(x, r)$ defined by (232) satisfy the engulfing property, as we show in the next lemma.

LEMMA 94. *Let (225) hold and suppose each $a_j(x)$ satisfies the doubling condition in each variable $x_i, i \neq j$, and is Lipschitz continuous in $x_2, ..., x_n$ uniformly in x_1. There is a geometric constant r_1 and a geometric constant C which also* *depends on the constants in the other conditions so that if $B(x,r) \cap B(y,r) \neq \phi$ and $0 < r < r_1$, then $B(y,r) \subset B(x, Cr)$.*

PROOF. Let $x = 0$ and choose a point $z \in B(0,r) \cap B(y,r)$. Then $|z_1| < r, |z_1 - y_1| < r, |z_i| < B_i(r)$ for $i \geq 2$, and $|z_i - y_i| < B_i(y,r)$ for $i \geq 2$. Let $\xi \in B(y,r)$. Then $|\xi_1 - y_1| < r$ and $|\xi_i - y_i| < B_i(y,r)$ for $i \geq 2$, and in order to prove that $\xi \in B(0, Cr)$, we must show that $|\xi_1| < Cr$ and $|\xi_i| < B_i(Cr), i \geq 2$. First,

$$|\xi_1| \leq |\xi_1 - y_1| + |y_1 - z_1| + |z_1| < r + r + r = 3r.$$

Next, for $i \geq 2$,

$$|\xi_i| \leq |\xi_i - y_i| + |y_i - z_i| + |z_i| < B_i(y,r) + B_i(y,r) + B_i(r).$$

But $B_i(y,r) \leq cB_i(r)$; in fact by Lemma 93, $B_i(y,r) \approx B_i(z,r) \approx B_i(r)$ since $z \in B(y,r)$ and $z \in B(0,r)$. Thus

$$|\xi_i| < c'B_i(r) \leq B_i(Cr), \quad i \geq 2,$$

and the lemma follows.

The flag balls form a δ-local prehomogeneous space. Lemma 94 yields the δ-local engulfing property (68) for the rectangles $B(x,r)$ with $\gamma = C$. Lemma 92 yields the δ-local doubling property (71) with $C_{doub} = 2^{d^* n} C^n$, and the δ-local weak monotonicity property (70). The δ-local scale property (69) is immediate, and the rectangles $B(x,r)$ are nonempty and open by our nondegeneracy assumption that $B_j(x,r) > 0$ when $r > 0$. This completes the demonstration that the family of flag balls $\mathcal{B} = \{B(x,r)\}_{x \in \mathbb{R}^n, 0 < r < \infty}$ forms a δ-local prehomogeneous space $\mathbb{B} = (\Omega, \mathcal{B})$ on Ω.

2.2. A_∞ properties of flag balls. Our next goal is to show that Definition 78 holds for the flag balls $B(x,r)$. However, the $A_j(x,r)$ may now vanish. This possibility is offset by the flag condition 13, which by Remark 87 implies that $B_j(x,r) > 0$ if $r > 0$. We also assume that (225) holds, together with an extra condition of A_∞ type which turns out to hold automatically if each $a_j(z)$ belongs to RH_∞ in each variable z_i, $i \neq j, 1$, uniformly in the other variables. In order to define this extra condition, fix x and r, and suppose the $B_j(x,r)$ are defined in the order $\{j_k\}_{k=2}^n$. For each j_k with $k \geq 3$, define

$$B^{(j_k)}(x,r) = \{z : |z_i - x_i| \leq B_i(x,r) \text{ if } i = j_2, \ldots, j_{k-1}; z_i = x_i \text{ if } i = 1, j_k, \ldots, j_n\}.$$

If we think of $B^{(j_k)}(x,r)$ as a $(k-2)$-dimensional rectangle in the variables $z_{j_2}, \ldots, z_{j_{k-1}}$, then its measure is given by

$$|B^{(j_k)}(x,r)|_{(k-2)} = \prod_{i=2}^{k-1} 2B_{j_i}(x,r).$$

We will assume the following A_∞-type condition: there are positive constants c and η such that for all $k \geq 3$, all x and r, and every measurable set $\mathcal{E} \subset B^{(j_k)}(x,r)$,

(233)

$$\frac{|\mathcal{E}|_{(k-2)}}{|B^{(j_k)}(x,r)|_{(k-2)}} \leq c \left(\frac{\int_{\mathcal{E}} A_{j_k}(z,r) \, dz_{j_2} \ldots dz_{j_{k-1}} |_{z_i = x_i, i = 1, j_k, \ldots, j_n}}{\int_{B^{(j_k)}(x,r)} A_{j_k}(z,r) \, dz_{j_2} \ldots dz_{j_{k-1}} |_{z_i = x_i, i = 1, j_k, \ldots, j_n}} \right)^\eta.$$

Condition (233) differs from ordinary A_∞ conditions because as x and r vary, so do the integrands on the right. An important feature of (233) is that when considered as a condition on any particular A_j, it does not involve the corresponding j-th coordinate (or the first coordinate). In practice, the condition is not easy to verify, but a more familiar condition which implies it is that each $a_j(z)$ satisfies the A_∞ condition in each $z_i, i \neq j, 1$, uniformly in the other variables, i.e., that there exist constants $c_1, \eta_1 > 0$ so that if I is a one-dimensional interval and $\mathcal{E}$ is a measurable subset of I, then

$$(234) \qquad \int_{\mathcal{E}} a_j(z) \, dz_i \leq c_1 \left(\frac{|\mathcal{E}|}{|I|} \right)^{\eta_1} \int_I a_j(z) \, dz_i, \quad i \neq j, 1,$$

for all $z_k, k \neq i$. In fact, by integrating both sides of this inequality with respect to z_1 and applying Fubini's Theorem, we obtain

$$\int_{\mathcal{E}} A_j(z, r) \, dz_i \leq c_1 \left(\frac{|\mathcal{E}|}{|I|} \right)^{\eta_1} \int_I A_j(z, r) \, dz_i, \quad i \neq j, 1,$$

for all r and $z_k, k \neq i$. As a consequence, $A_j(z, r)$ belongs to product A_∞ (i.e., A_∞ for arbitrary rectangles with sides parallel to the coordinate axes) in any subcollection of the variables $z_2, \ldots, z_{j-1}, z_{j+1}, \ldots, z_n$ uniformly in r, z_1, z_j and the variables which are not in the subcollection. From this, we obtain (233) as a corollary, and in fact we even obtain a stronger statement: there are constants $c, \eta > 0$ such that if $3 \leq k \leq n$ and $z_{i_2}, \ldots, z_{i_{k-1}}$ is any subcollection of $z_2, \ldots, z_{j-1}, z_{j+1}, \ldots, z_n$, R is any $(k-2)$-dimensional rectangle with edges parallel to the coordinate axes, and $\mathcal{E}$ is a measurable subset of R, then

$$\frac{|\mathcal{E}|_{(k-2)}}{|R|_{(k-2)}} \leq c \left(\frac{\int_{\mathcal{E}} A_j(z, r) \, dz_{i_2} \ldots dz_{i_{k-1}}}{\int_R A_j(z, r) \, dz_{i_2} \ldots dz_{i_{k-1}}} \right)^{\eta}$$

for all $r > 0$ and $z_\ell \notin \{z_{i_2}, \ldots, z_{i_{k-1}}\}, 1 \leq \ell \leq n$.

We note in passing that since RH_∞ implies A_∞, (234) (and so also (233)) holds if each $a_j(z)$ satisfies RH_∞ in each variable $z_i, i \neq j, 1$, uniformly in the other variables. Thus the hypotheses of Theorem 17 yield condition (233).

We begin by showing that for most z in a flag ball $B(x, r)$, $A_j(z, r)$ is almost as large as its average over the flag ball $B(x, r)$.

LEMMA 95. *Suppose that (225) and (233) hold, and let $\varepsilon > 0$. Then there are positive constants r_0 and c so that for all x and all r with $0 < r \leq r_0$,*

$$\left| \left\{ z \in B(x, r) : A_j(z, r) > \varepsilon \frac{1}{|B(x, r)|} \int_{B(x, r)} A_j(y, r) \, dy \right\} \right| \geq (1 - c\varepsilon^{\frac{\eta}{1-\eta}}) |B(x, r)|.$$

Here, r_0 and c are geometric constants which also depend on the constants in the other conditions, and r_0 also depends on ε. The value of η is the same as in (233).

Thus, if ε is small and $r \leq r_0$, $A_j(\cdot, r)$ exceeds ε times the average over $B(x, r)$ of $A_j(\cdot, r)$ on a large portion of $B(x, r)$.

PROOF. Fix x and r, and write $B = B(x, r)$, etc. We consider the case $x = 0$ with the natural ordering of the $B_j(r)$. The conclusion in case $j = 2$ is obvious by (229), so we may consider fixed $j \geq 3$. Let

$$\mathcal{E} = \left\{ \zeta \in B^{(j)} : A_j(\zeta, r) > \varepsilon \frac{1}{|B^{(j)}|_{(j-2)}} \right.$$
$$\left. \times \int_{B^{(j)}} A_j((0, y_2, \ldots, y_{j-1}, 0, \ldots, 0), r) \, dy_2 \ldots dy_{j-1} \right\},$$

where
$$B^{(j)} = \{\zeta = (0, \zeta_2, \ldots, \zeta_{j-1}, 0, \ldots, 0) : |\zeta_i| \le B_i(r)\}.$$
We first show that
$$|\mathcal{E}|_{(j-2)} \ge (1 - c\varepsilon^{\frac{\eta}{1-\eta}})|B^{(j)}|_{(j-2)}.$$
Indeed, by (233), there exist $c, \eta > 0$ such that
$$\frac{|B^{(j)} \setminus \mathcal{E}|_{(j-2)}}{|B^{(j)}|_{(j-2)}} \le c \left(\frac{\int_{B^{(j)} \setminus \mathcal{E}} A_j((0, \zeta_2, \ldots, \zeta_{j-1}, 0, \ldots, 0), r) \, d\zeta_2 \ldots d\zeta_{j-1}}{\int_{B^{(j)}} A_j((0, \zeta_2, \ldots, \zeta_{j-1}, 0, \ldots, 0), r) \, d\zeta_2 \ldots d\zeta_{j-1}} \right)^\eta$$
$$\le c \left(\frac{\varepsilon |B^{(j)} \setminus \mathcal{E}|_{(j-2)}}{|B^{(j)}|_{(j-2)}} \right)^\eta$$

by definition of $\mathcal{E}$. Thus, $|B^{(j)} \setminus \mathcal{E}|_{(j-2)} \le c\varepsilon^{\frac{\eta}{1-\eta}}|B^{(j)}|_{(j-2)}$, and consequently,
$$|\mathcal{E}|_{(j-2)} = |B^{(j)}|_{(j-2)} - |B^{(j)} \setminus \mathcal{E}|_{(j-2)}$$
$$\ge |B^{(j)}|_{(j-2)} - c\varepsilon^{\frac{\eta}{1-\eta}}|B^{(j)}|_{(j-2)} = (1 - c\varepsilon^{\frac{\eta}{1-\eta}})|B^{(j)}|_{(j-2)},$$
as desired. Next, let $\zeta \in \mathcal{E}$, $\zeta = (0, \zeta_2, \ldots, \zeta_{j-1}, 0, \ldots, 0)$, and consider any $z \in B$ with $z_i = \zeta_i$ for $i = 2, \ldots, j-1$ and $z_1, z_j, \ldots, z_n$ unrestricted. By doubling in the first variable,
$$\frac{1}{c_1} A_j(z, r) \ge A_j((0, z_2, \ldots, z_n), r)$$
$$\ge A_j(\zeta, r) - |A_j((0, z_2, \ldots, z_n), r) - A_j(\zeta, r)|$$
$$\ge A_j(\zeta, r) - \int_0^r \left(\sum_{i=j}^n C_L |z_i| \right) dt \qquad \text{by definition of } z, \zeta$$
$$\ge A_j(\zeta, r) - rC_L \sum_{i=j}^n CB_i(r) \qquad \text{since } z \in B$$
$$\ge A_j(\zeta, r) - rC_L(n-2)(1 + rC_L)^{n-2}CB_j(r) \quad \text{by Lemma 88.}$$
Since $\zeta \in \mathcal{E}$,
$$A_j(\zeta, r) > \varepsilon \frac{1}{|B^{(j)}|_{(j-2)}} \int_{B^{(j)}} A_j((0, y_2, \ldots, y_{j-1}, 0, \ldots, 0), r) \, dy_2 \ldots dy_{j-1}$$
$$\ge \varepsilon c_2 B_j(r)$$
by Remark 91. Combining estimates, we obtain
$$\frac{1}{c_1} A_j(z, r) \ge \left[\varepsilon c_2 - rC_L(n-2)(1 + rC_L)^{n-2}C \right] B_j(r)$$
$$\ge \frac{1}{2}\varepsilon c_2 B_j(r)$$
if $r \le r_1$ for a sufficiently small constant r_1 depending on ε. Hence, by (228),
$$A_j(z, r) \ge \frac{1}{2}c_1 c_2 \varepsilon c_3 \frac{1}{|B|} \int_B A_j(y, r) \, dy$$
if z belongs to the Cartesian product
$$(-r, r) \times \mathcal{E} \times \prod_{i=j}^n (-B_i(r), B_i(r)).$$

The measure of this product is

$$|\mathcal{E}|_{(j-2)} 2r \prod_{i=j}^{n} 2B_i\left(r\right) \geq (1 - c\varepsilon^{\frac{\eta}{1-\eta}})|B^{(j)}|_{(j-2)} 2r \prod_{i=j}^{n} 2B_i\left(r\right) = (1 - c\varepsilon^{\frac{\eta}{1-\eta}})|B|,$$

which proves the lemma.

REMARK 96. *The conclusion of Lemma 95 holds if instead of assuming (225) and (233), we assume only that there are positive constants c and η independent of x and r such that for each j and every measurable set $\mathcal{E}$ with $\mathcal{E} \subset B\left(x,r\right)$,*

$$\frac{|\mathcal{E}|}{|B\left(x,r\right)|} \leq c \left(\frac{\int_{\mathcal{E}} A_j\left(z,r\right) dz}{\int_{B(x,r)} A_j\left(z,r\right) dz} \right)^{\eta}.$$

This can be proved in a simpler way than Lemma 95, without considering Cartesian products, and we leave the verification to the reader. The condition holds if $a_j\left(z\right)$ satisfies the A_∞ condition in each variable x_i uniformly in the other variables. However, as mentioned earlier, an advantage of (233) is that it requires no hypothesis on a_j in the j-th variable.

LEMMA 97. *Let (225) and (233) hold. Then given $\nu > 0$, there exists $\varepsilon > 0$ independent of x and r such that*

$$|\{z \in B\left(x,r\right) : A_j\left(z,r\right) > \varepsilon B_j\left(x,r\right) \quad \text{for all } j\}| \geq (1-\nu)\,|B\left(x,r\right)|.$$

PROOF. It is enough to prove the inequality for each fixed j. Denote $B = B\left(x,r\right)$. By Lemma (95), there exist $c, \eta > 0$ such that for all $\varepsilon > 0$,

$$\left| \left\{ z \in B : A_j\left(z,r\right) > \varepsilon \frac{1}{|B|} \int_B A_j\left(y,r\right) dy \right\} \right| \geq (1 - c\varepsilon^{\eta})|B|.$$

Therefore, by (225) and Lemma 90,

$$|\{z \in B : A_j\left(z,r\right) > \varepsilon B_j\left(x,r\right)\}| \geq (1 - c'\varepsilon^{\eta})|B|,$$

and we're done.

Adapted flag balls. We have already shown at the end of subsection 5.2.1 that the flag balls $\mathcal{B}$ form a δ-local prehomogeneous space $\mathbb{B} = (\Omega, \mathcal{B})$ on Ω, and Lemma 89 yields the size-limiting condition (91). The hypotheses of Theorems 17 and 23 imply both (225) and (233). Thus in Lemmas 89 and 97 we have established conditions 1 and 2 respectively of Definition 78, and in Lemma 92 we have established the reverse doubling condition (215). As mentioned earler, this completes the proofs of Theorems 17 and 23.

REMARK 98. *Under the hypotheses of Theorem 17 or 23, the ratio $\frac{d(x,x')}{\rho}$ in the Hölder estimate (175) of Theorem 65 can be replaced with the larger quantity*

$$\max \left\{ \frac{|x_1 - x_1'|}{\rho}, \left(\frac{|x_2 - x_2'|}{B_2\left(y,\rho\right)} \right)^{\frac{1}{d^*}}, ..., \left(\frac{|x_n - x_n'|}{B_n\left(y,\rho\right)} \right)^{\frac{1}{d^*}} \right\},$$

where d^* is the doubling exponent in (230). Indeed, the quasimetric $d(x,x')$ in (175) is the symmetric quasimetric appearing in Theorem 8, which in turn arises from an application of Lemma 37 to the quasimetric associated in Proposition 41 to the extension of the δ-local prehomogeneous space $\mathbb{B}$. The above estimate now follows from a calculation using the definition (59) of quasimetric, the weak monotonicity (70) of the flag balls in $\mathbb{B}$, and the second inequality in (230).

CHAPTER 6

Appendix

Here we collect some technical results needed for our development above, as well as some peripheral material mentioned above. In subsection 6.1 we demonstrate the necessity of the Fefferman-Phong condition for L^q-subellipticity. In subsection 6.2 we show the near necessity of the Sobolev and Poincaré inequalities for some related notions of subellipticity for Dirichlet and Neumann boundary value problems. In subsection 6.3 we show that the noninterference balls $A(x,r)$ associated to a collection of vector fields $\mathcal{X}$ are Ω-locally equivalent to the subunit balls $K(x,r)$ if and only if the vector fields are reverse Hölder of infinite order in the first variable x_1. Subsection 6.4 exhibits a collection $\mathcal{X}$ of vector fields such that Theorem 82 or 8 applies to the associated family of noninterference balls $A(x,r)$ to show that $\mathcal{X}$ is L^q-subelliptic, but for which the noninterference balls $A(x,r)$ fail to be Ω-locally equivalent to the subunit balls $K(x,r)$. Subsection 6.5 contains the technical proof of a product reverse Hölder implication needed in section 5. Subsection 6.6 demonstrates some implications regarding the noninterference condition and its strong form. In the final two subsections, 6.7 and 6.8, we briefly describe some alternate notions of weak solution to a degenerate elliptic equation, and some alternate methods of proof.

1. Necessity of the Fefferman-Phong condition

In subsection 1.1.1 of the introduction we stated that the Fefferman-Phong condition (5) is necessary for the stable subellipticity of the continuous quadratic form $\mathcal{Q}(x,\xi)$. Here is the proof.

PROPOSITION 99. *Let $\mathcal{Q}(x,\xi)$ be a nonnegative semidefinite continuous quadratic form in Ω. Suppose that the family of quadratic forms $\mathcal{Q}_\tau(x,\xi) = \mathcal{Q}(x,\xi) + \tau |\xi|^2$, $0 < \tau < 1$, is subelliptic in Ω in the sense of Definition 7 uniformly in $0 < \tau < 1$, in the sense that (25) holds with α and $\mathcal{C}$ independent of $0 < \tau < 1$. Then the containment condition (19) holds.*

PROOF. Let $\mathcal{Q}(x,\xi) = \xi'Q(x)\xi$ with the matrix $Q(x)$ symmetric. If $P_\tau(x)$ is a measurable matrix with $P_\tau(x)' P_\tau(x) = Q(x) + \tau I$, then with y fixed and $u_\tau(x) = \delta^\tau(x,y)$, the fundamental inequality (21), or more precisely (176), shows that

$$L_\tau u_\tau = T'g$$

where

$$L_\tau = \nabla (Q(x) + \tau I) \nabla = [P_\tau(x)\nabla]' [P_\tau(x)\nabla]$$

and

$$g = P_\tau(x)\nabla u \text{ is bounded independent of } 0 < \tau < 1 \text{ by } (176),$$
$$T = P_\tau(x)\nabla \text{ is subunit with respect to } Q(x) + \tau I.$$

133

The uniform L^∞-subellipticity of $\mathcal{Q}(x,\xi) + \tau |\xi|^2$ then shows that u_τ is Hölder continuous of order $\alpha > 0$, independent of $0 < \tau < 1$, and it follows that

$$\delta^\tau (x,y) = |u_\tau(x) - u_\tau(y)| \leq C |x-y|^\alpha,$$

with C and α independent of $0 < \tau < 1$. Lemma 66 now shows that

$$\delta(x,y) = \lim_{\tau \to 0} \delta^\tau(x,y) \leq C |x-y|^\alpha,$$

which yields (5) with $\varepsilon = \alpha$.

In certain cases when $\mathcal{Q}$ is a sum of squares, we can show the necessity of the ε-comparability condition (5) for subellipticity, rather than stable subellipticity, of $\mathcal{Q}$.

PROPOSITION 100. *Suppose the vector fields $X_j = a_j(x) \frac{\partial}{\partial x_j}$, $1 \leq j \leq n$, are Lipschitz continuous and satisfy the subelliptic Definition 11 in Ω. Then there is $\varepsilon > 0$ such that the ε-comparability condition (5) holds for the subunit balls $K(x,r)$ associated to the quadratic form $\mathcal{Q}(x,\xi) = \sum_{j=1}^n (a_j(x)\xi_j)^2$.*

PROOF. Suppose the vector fields $\{X_j\}_{j=1}^n$ are subelliptic, and let L be the "sum of squares" operator $\sum_{j=1}^n X_j' X_j$. Then by (3), the quadratic form inequality (33) holds with $c_{sym} = C_{sym} = 1$. Let α be the positive Hölder exponent that arises in Definition 11 for a compact neighbourhood of x, and set $N = \frac{1}{\alpha}$. We will first show the following inequality by refining the argument used in Proposition 15 of the introduction to prove the necessity of the flag condition:

$$(235) \qquad \sup_{|z_i - x_i| \leq r\chi_{\mathcal{I}}(i)} \sum_{j \notin \mathcal{I}} a_j(z) \geq cr^N, \qquad 0 < r < r_0, \phi \neq \mathcal{I} \subsetneq \{1,2,...,n\},$$

for a constant c independent of x and r. By reordering variables, we suppose $\mathcal{I} = \{m+1, m+2, ..., n\}$ for some $1 \leq m < n$. As in the proof of Proposition 15, define φ on $[0,\infty)$ by $\varphi(t) = \begin{cases} t, & 0 \leq t \leq 1 \\ 1, & t > 1 \end{cases}$, and let $\psi(t)$ be a smooth compactly supported function in $(-1,1)$ that is identically one on $\left(-\frac{1}{2}, \frac{1}{2}\right)$. Fix x and $0 < r < 1$, and suppose that

$$\delta = r^{-N} \sup_{|z_i - x_i| \leq r\chi_{\mathcal{I}}(i)} \sum_{j=1}^m a_j(z) < 1.$$

Set $u(z) = r\varphi\left(\frac{|z' - x'|}{\delta r^N}\right) \psi\left(\frac{|z'' - x''|}{r}\right)$ where $z' = (z_1, ..., z_m)$ and $z'' = (z_{m+1}, ..., z_n)$, and define $f_j = X_j u = a_j \frac{\partial u}{\partial x_j}$. Since a_j is Lipschitz and satisfies $a_j(z) \leq \delta r^N$ on $\{z : |z_i - x_i| \leq r\chi_{\mathcal{I}}(i)\}$ for $1 \leq j \leq m$, we have $|a_j(z)| \leq \delta r^N + C|z' - x'|$ for $z \in B_r \equiv \prod_{i=1}^n [x_i - r, x_i + r]$, and so then

$$|f_j(z)| = \left| a_j(z) \frac{\partial u}{\partial x_j}(z) \right| \leq Cra_j(z) \left| \varphi'\left(\frac{|z' - x'|}{\delta r^N}\right) \right| \frac{1}{\delta r^N}$$

$$\leq Cr\left(\delta r^N + C\delta r^N\right) \frac{1}{\delta r^N} \leq C,$$

for $1 \leq j \leq m$. For $j > m$, we have

$$|f_j(z)| = \left| a_j(z) \frac{\partial u}{\partial x_j}(z) \right| \leq Cr \left| \psi'\left(\frac{|z'' - x''|}{r}\right) \right| \frac{1}{r} \leq C.$$

Thus u is a $W^{1,2}(\Omega)$ weak solution of

$$Lu = \sum_{j=1}^{n} T_j' f_j,$$

where $T_j = X_j$ is subunit, and $\|u\|_2, \|f_j\|_\infty \le C$, and so $\|u\|_{Lip_\alpha} \le C$ by the subellipticity of $\{X_j\}_{j=1}^n$. However, if $|z' - x'| = \delta r^N$, then

$$
\begin{aligned}
C &\ge \|u\|_{Lip_\alpha} \ge \frac{u(z', x'') - u(x', x'')}{|z' - x'|^\alpha} = \frac{r\varphi(1)\psi(0) - r\varphi(0)\psi(0)}{|z' - x'|^\alpha} \\
&= \frac{r - 0}{|z' - x'|^\alpha} = \delta^{-\alpha} r^{1 - N\alpha} = \delta^{-\alpha},
\end{aligned}
$$

which yields (235) with $c = C^{-\frac{1}{\alpha}}$.

We will now obtain the Fefferman-Phong condition (5) with $\varepsilon = \alpha^n$, by using (235) to construct a subunit curve of length Cr connecting an arbitrary pair of points x, y with $|x - y| \le r^{N^n}$. Recall $N = \frac{1}{\alpha}$. So fix x, y with $|x - y| \le r^{N^n}$, and without loss of generality we may assume that $a_1(x) \equiv 1$ and that $y_j \ge x_j$ for $1 \le j \le n$. Now by (235) with $\mathcal{I} = \{1\}$, and reordering variables, there is z_1^1 in $[x_1 - r, x_1 + r]$ with $a_2(z_1^1, x_2, ..., x_n) \ge cr^N$. Let $r_2 = \beta c r^{N+1} < \beta c r^N$ with β sufficiently small (depending only on the Lipschitz constant) that $a_2(w) \ge \frac{c}{2} r^N$ for $|w - (z_1^1, x_2, ..., x_n)| \le 100 r_2$. By (235) with $\mathcal{I} = \{1, 2\}$ and $r = r_2$, and reordering variables, there is (z_1^2, z_2^2) in $[z_1^1 - r_2, z_1^1 + r_2] \times [x_2 - r_2, x_2 + r_2]$ with $a_3(z_1^2, z_2^2, x_3, ..., x_n) \ge c r_2^N = c_3 r^{N^2 + N}$. Then with $r_3 = \beta c_3 r^{N^2 + N + 1}$ (same β), we have $a_3(w) \ge \frac{c_3}{2} r^{N^2 + N}$ for $|w - (z_1^2, z_2^2, x_3, ..., x_n)| \le 100 r_3$. Continuing in this way we obtain $(z_1^{j-1}, ..., z_{j-1}^{j-1})$ in $\prod_{i=1}^{j-2} [z_i^{j-2} - r_{j-1}, z_i^{j-2} + r_{j-1}] \times [x_{j-1} - r_{j-1}, x_{j-1} + r_{j-1}]$, $3 \le j \le n$, and $r_2, ..., r_n$ such that with $N_j = \sum_{i=0}^{j-1} N^i$,

$$a_j\left(z_1^{j-1}, ..., z_{j-1}^{j-1}, x_j, ..., x_n\right) \ge c_j r^{N_j}, \quad 2 \le j \le n,$$

and

$$(236) \qquad a_j(w) \ge \frac{c_j}{2} r^{N_j} \quad \text{if} \quad \left|w - \left(z_1^{j-1}, ..., z_{j-1}^{j-1}, x_j, ..., x_n\right)\right| \le 100 r_j,$$

where $r_j = \beta c_j r^{N_j}$, and the constants c_j depend only on β, N and c in (235), $2 \le j \le n$.

We can now connect $(x_1, ..., x_n)$ to $(z_1^1, x_2, ..., x_n)$ by a subunit curve - γ is subunit if $(\gamma'(t) \cdot \xi)^2 \le \sum_{j=1}^n a_j(\gamma(t))^2 \xi_j^2$ - of length at most r, namely

$$\gamma(t) = (x_1 + t, x_2, ..., x_n), \quad 0 \le t \le z_1^1 - x_1.$$

Then using (236) with $j = 2$ we connect $(z_1^1, x_2, ..., x_n)$ to $(z_1^2, z_2^2, x_3, ..., x_n)$ by a subunit curve consisting of two segments each of length at most r, namely

$$
\begin{aligned}
\gamma_1(t) &= (z_1^1 + t, x_2, ..., x_n), && 0 \le t \le z_1^2 - z_1^1 \le r_2 \le r, \\
\gamma_2(t) &= \left(z_1^2, x_2 + \frac{c_2}{2} r^N t, x_3, ..., x_n\right), && 0 \le t \le 2\frac{z_2^2 - x_2}{c_2 r^N} \le r,
\end{aligned}
$$

where we have used that $\frac{2r_2}{c_2 r^N} \le r$ and $c_2 = c$. Using (236) with $j = 3$ we similarly connect $(z_1^2, z_2^2, x_3, ..., x_n)$ to $(z_1^3, z_2^3, z_3^3, x_4, ..., x_n)$ by a subunit curve consisting of three segments each of length at most r. We continue this process until we reach $(z_1^{n-1}, ..., z_{n-1}^{n-1}, x_n)$. Then using (236) with $j = n$ and $|x - y| \le r^{N_n}$

(recall that $N_n \leq N^n$, $r < 1$ and that we initially assumed $|x - y| \leq r^{N_n}$) we can connect $\left(z_1^{n-1}, ..., z_{n-1}^{n-1}, x_n\right)$ to $\left(z_1^{n-1}, ..., z_{n-1}^{n-1}, y_n\right)$, then $\left(z_1^{n-1}, ..., z_{n-1}^{n-1}, y_n\right)$ to $\left(z_1^{n-2}, ..., z_{n-2}^{n-2}, y_{n-1}, y_n\right)$, etc., until we finally connect $\left(z_1^1, y_2, ..., y_n\right)$ to $(y_1, ..., y_n)$, all by subunit curves of multiple segments and lengths at most Cr. Since $N_n \leq N^n$, this completes the proof of the proposition.

REMARK 101. *From the above proof, we also obtain that subellipticity of Lipschitz vector fields $\{X_j\}_{j=1}^n$, $X_j = a_j(x)\frac{\partial}{\partial x_j}$, implies the ε-comparability condition for the flag balls $B(x,r)$ in place of $K(x,r)$:*

$$D(x,r) \subset B(x, Cr^\varepsilon), \qquad x \in \Omega, 0 < r \leq 1.$$

Of course, the remark is a direct corollary of Proposition 100 if $\mathcal{K} \subset \mathcal{B}$, as is the case if the size limiting condition (91) holds and the a_j are reverse Hölder in the x_1 variable, uniformly in the other variables.

2. Necessity of the Sobolev and Poincaré inequalities

As stated in subsection 1.1.1 of the introduction, the Sobolev inequality (15) is necessary for a variant of the notion of subellipticity of the quadratic form $\mathcal{Q}$. Recall that $\mathcal{Q}(x,\xi) = \xi'Q(x)\xi$ is L^q-subelliptic relative to the homogeneous Dirichlet problem for the balls $B(x,r)$, if we assume existence of weak solutions u to homogeneous Dirichlet problems for the balls B:

$$(237) \qquad \begin{cases} Lu &= f \quad in\ B \\ u &= 0 \quad on\ \partial B \end{cases},$$

where $L = \nabla'Q(x)\nabla$ and $f \in L^{\frac{q}{2}}(B)$; as well as the global boundedness estimate,

$$(238) \qquad \sup_{z \in B} |u(z)| \leq C \left(\int_B |f|^{\frac{q}{2}}\right)^{\frac{2}{q}},$$

for these weak solutions u to (237). We will restrict our statement and proof to the case where weak solution here is interpreted in the classical $W^{1,2}$ sense, although the proof adapts readily to other notions of weak solution.

LEMMA 102. *Suppose that $\mathcal{Q}(x,\xi)$ is L^q-subelliptic relative to the homogeneous Dirichlet problem for the balls $B(x,r)$ in the classical $W^{1,2}$ sense for some $q < Q_*$, where Q_* is the lower dimension of the space of balls $B(x,r)$. Then the Sobolev inequality (15) holds with $\sigma = \frac{q}{q-2}$.*

PROOF. Given $f \in L^{\frac{q}{2}}(B)$ with $q = 2\sigma' < Q_*$, let u be the weak $W^{1,2}$ solution to the degenerate elliptic Dirichlet problem (237) where $L = \nabla'Q(x)\nabla$ is the operator with symbol $\xi'Q(x)\xi = \mathcal{Q}(x,\xi)$. Then for $w \in Lip_0(B)$ and f nonnegative, we have

$$\int_B w^2 f = \int_B w^2 \nabla'Q(x)\nabla u = -2\int_B w \langle \nabla u, \nabla w \rangle$$
$$\leq \left\{\int_B w^2 \|\nabla u\|_\mathcal{Q}^2\right\}^{\frac{1}{2}} \left\{\int_B \|\nabla w\|_\mathcal{Q}^2\right\}^{\frac{1}{2}}.$$

Now the square of the first factor satisfies

$$
\begin{aligned}
\int_B w^2 \|\nabla u\|_Q^2 &= \int_B (\nabla u)' \, w^2 Q\,(x) \, \nabla u = -\int_B u \nabla' \left[w^2 Q\,(x) \, \nabla u \right] \\
&= -2 \int_B \langle u \nabla w, w \nabla u \rangle - \int_B u w^2 f \\
&\leq \frac{1}{2} \int_B w^2 \|\nabla u\|_Q^2 + 2 \int_B u^2 \|\nabla w\|_Q^2 + \int_B |u|\, w^2 f,
\end{aligned}
$$

and absorbing the first term on the right-hand side yields

$$
\begin{aligned}
(239) \int_B w^2 \|\nabla u\|_Q^2 &\leq 4 \left(\sup_B |u| \right)^2 \int_B \|\nabla w\|_Q^2 + 2 \left(\sup_B |u| \right) \int_B w^2 f \\
&\leq C \max \left\{ \left(\sup_B |u| \right)^2 \int_B \|\nabla w\|_Q^2 , \left(\sup_B |u| \right) \int_B w^2 f \right\}.
\end{aligned}
$$

No matter which term on the right-hand side of (239) is actually the maximum, these inequalities together with (238) give

$$
\begin{aligned}
\int_B w^2 f &\leq C \left(\sup_B |u| \right) \int_B \|\nabla w\|_Q^2 \\
&\leq C \left(\int_B f^{\frac{q}{2}} \right)^{\frac{2}{q}} \int_B \|\nabla w\|_Q^2 \\
&= C |B|^{\frac{1}{\sigma'}} \left(\frac{1}{|B|} \int_B f^{\sigma'} \right)^{\frac{1}{\sigma'}} \int_B \|\nabla w\|_Q^2 ,
\end{aligned}
$$

Thus we have

$$
\left(\frac{1}{|B|} \int_B |w|^{2\sigma} \right)^{\frac{1}{\sigma}} = \sup_{\frac{1}{|B|} \int_B |f|^{\sigma'} = 1} \left| \frac{1}{|B|} \int_B w^2 f \right| \leq C |B|^{\frac{1}{\sigma'}} \left(\frac{1}{|B|} \int_B \|\nabla w\|_Q^2 \right).
$$

Since $q < Q_*$, we have $|B| = |B\,(y,r)| \leq C r^q$, and so we obtain

$$
\left(\frac{1}{|B|} \int_B |w|^{2\sigma} \right)^{\frac{1}{2\sigma}} \leq C r^{\frac{q}{2\sigma'}} \left(\frac{1}{|B|} \int_B \|\nabla w\|_Q^2 \right)^{\frac{1}{2}},
$$

for all $w \in Lip_0\,(B)$. Using the density of $Lip_0\,(B)$ in $W_0^{1,2}\,(B)$, we finally obtain (15) with $\sigma = \frac{q}{q-2}$ since $q = 2\sigma'$.

Now recall that $Q\,(x,\xi)$ is L^2-hypoelliptic relative to the homogeneous Neumann problem for the balls $B\,(x,r)$, if we assume existence of weak solutions to the homogeneous Neumann problem for the balls $B = B\,(x,r)$,

$$
(240) \qquad \begin{cases} Lu = f & in\ B \\ \mathbf{n}_Q u = 0 & on\ \partial B \end{cases},
$$

where $L = \nabla' Q\,(x)\, \nabla$, $\mathbf{n}_Q = \mathbf{n}' B\,(x)\, \nabla$ with $\mathbf{n}$ the unit outward normal to ∂B, and $f \in W^{1,2}\,(B)$ satisfies the compatibility condition $\int_B f = 0$; and if we also assume the natural hypoelliptic estimate

$$
(241) \qquad \|u\|_{L^2(B)} \leq C r^2 \|f\|_{L^2(B)},
$$

for these weak solutions u to the above Neumann problem. Again, we will restrict our statement and proof to the case where weak solution here is interpreted in the classical $W^{1,2}$ sense, although the proof adapts readily to other notions of weak solution.

LEMMA 103. *Suppose that $Q(x, \xi)$ is L^2-hypoelliptic relative to the homogeneous Neumann problem for the balls $B(x, r)$ in the classical $W^{1,2}$ sense. Then the Poincaré inequality (17) holds.*

PROOF. Recall that $u \in W^{1,2}(B)$ is a weak solution of the boundary value problem (240) if (31) holds, i.e.

$$(242) \qquad -\int_B (\nabla v)' Q(x) \nabla u = \int_B vf, \quad \text{for all } v \in W^{1,2}(B).$$

Now given $f \in W^{1,2}(B)$, let u be the weak $W^{1,2}$ solution to the homogeneous Neumann problem

$$(243) \qquad \begin{cases} Lu & = & f - f_B & in\ B \\ \mathbf{n}_Q u & = & 0 & on\ \partial B \end{cases},$$

where $f_B = \frac{1}{|B|} \int_B f$ is the average of f on B. With $v = u$ in (242), we obtain

$$\begin{aligned} \int_B Q(x, \nabla u) & = & \int_B (\nabla u)' Q(x) \nabla u \\ & = & -\int_B u(f - f_B), \end{aligned}$$

and so

$$(244) \qquad \int_B Q(x, \nabla u) \le \|u\|_{L^2(B)} \|f - f_B\|_{L^2(B)} \le Cr^2 \|f - f_B\|_{L^2(B)}^2$$

by (241) with f replaced by $f - f_B$. We now conclude using the vanishing mean value of f, (242) with $v = f - f_B$, the Cauchy-Schwarz inequality and finally (244), that

$$\begin{aligned} \|f - f_B\|_{L^2(B)}^2 & = & \int_B (f - f_B) f \\ & = & -\int_B (\nabla f)' Q(x) \nabla u \\ & \le & \left\{ \int_B Q(x, \nabla f) \right\}^{\frac{1}{2}} \left\{ \int_B Q(x, \nabla u) \right\}^{\frac{1}{2}} \\ & \le & \left\{ \int_B Q(x, \nabla f) \right\}^{\frac{1}{2}} \sqrt{C} r \|f - f_B\|_{L^2(B)}. \end{aligned}$$

Dividing through by $\|f - f_B\|_{L^2(B)}$ yields (17) as required.

3. The noninterference balls $A(x, r)$ and the reverse Hölder condition

Recall the family $\mathcal{A}$ of noninterference balls

$$A(x, r) = \prod_{j=1}^n (x_j - A_j(x, r), x_j + A_j(x, r))$$

where $A_j(x, r) = \int_0^r a_j(x_1 + t, x_2, ..., x_n) \, dt$, and the family $\mathcal{K}$ of subunit balls $K(x, r)$. Proposition 44 shows that under the hypotheses of Theorem 20, $\mathcal{A} \subset \mathcal{K}$ since $\mathcal{A}$ is a general homogeneous space and the family of balls $\mathcal{A}$ satisfies the subrepresentation inequality (90) with respect to the vector fields $\{X_j\}_{j=1}^n$, $X_j = a_j(x)\frac{\partial}{\partial x_j}$, and the balls $A(x,r)$. However, these properties hold under the following conditions weaker than those assumed in Theorem 20; the vector fields $X_1 = \frac{\partial}{\partial x_1}$, $X_j = a_j\frac{\partial}{\partial x_j}$, $2 \leq j \leq n$, are continuous, satisfy the noninterference Definition 18 and (40), and the coefficients a_j are Lipschitz continuous in $x_2, ..., x_n$ uniformly in x_1, and *merely doubling* in the first variable x_1 uniformly in $x_2, ..., x_n$. Indeed, we showed in subsection 5.1 that under these hypotheses, $\mathcal{A}$ is a homogeneous space adapted to the vector fields $\mathcal{X}$. Proposition 80 then yields the proportional subrepresentation inequality (182) for $\mathcal{A}$, and finally Lemma 72 yields the standard subrepresentation inequality (186).

Our next result shows that under these weaker hypotheses, the *reverse* implication $\mathcal{K} \subset \mathcal{A}$ holds if and only if the a_j are reverse Hölder of infinite order in x_1. The reader is reminded that while the noninterference Definition 18 is precisely the size limiting condition (91) for the noninterference balls $\mathcal{A}$, Proposition 44 does *not* imply that $\mathcal{K} \subset \mathcal{A}$, unless the a_j are also reverse Hölder in the x_1 variable, uniformly in the other variables.

PROPOSITION 104. *Let* $X_1 = \frac{\partial}{\partial x_1}$, $X_j = a_j\frac{\partial}{\partial x_j}$, $2 \leq j \leq n$, *be continuous vector fields satisfying the noninterference condition in Definition* 18, *and with* a_j *Lipschitz continuous in* $x_2, ..., x_n$ *uniformly in* x_1, *and doubling in* x_1 *uniformly in* $x_2, ..., x_n$. *Then the following are equivalent:*

(1) $\mathcal{A} \cong \mathcal{K} \cong \widetilde{\mathcal{K}} \cong \mathcal{K}^*$,
(2) $\mathcal{K} \subset \mathcal{A}$,
(3) a_j *is reverse Hölder of infinite order in* x_1 *for* $2 \leq j \leq n$.

PROOF. If a_j is reverse Hölder in x_1 for $2 \leq j \leq n$, then the family of balls $\mathcal{A}$ is equivalent to the family of subunit balls $\mathcal{K}$ by Proposition 44. Indeed, Definition 18 implies (91), and $\widetilde{\mathcal{B}} = \mathcal{B}$ since the balls $A(x,r)$ are rectangles. Remark 46 in subsection 2.3 then yields $\mathcal{A} \cong \mathcal{K} \cong \widetilde{\mathcal{K}} \cong \mathcal{K}^*$.

Conversely, if $\mathcal{K} \subset \mathcal{A}$, then since $A(x,r)$ is doubling in r, we have for some constant C_0,

$$K(x,r) \subset C_0 A(x,r)$$

for all $x \in \Omega$, $0 < r < r_0$, where $C_0 A(x,r)$ denotes the rectangle centered at x obtained by multiplying the side lengths of $A(x,r)$ by C_0. Now fix $x \in \Omega$ and $r > 0$. Choose $t_j \in [0, r]$ so that

$$\max_{0 \leq t \leq r} a_j(x_1 + t, x_2, ..., x_n) = a_j(x_1 + t_j, x_2, ..., x_n),$$

and assume, in order to derive a contradiction, that the third property fails, i.e. for some $2 \leq j \leq n$,

$$m_j \equiv a_j(x_1 + t_j, x_2, ..., x_n) \geq 2C_1 \frac{1}{r} \int_0^r a_j(x_1 + t, x_2, ..., x_n)\, dt = 2C_1 \frac{1}{r} A_j(x,r),$$

where C_1 is a sufficiently large multiple (related to the doubling assumption) of C_0 to be chosen later. Since a_j is Lipschitz in x_j with constant β_j, it follows that

(245) $$a_j(x_1 + t_j, x_2, ..., x_{j-1}, x_j + s, x_{j+1}, ..., x_n) \geq \frac{1}{2}m_j$$

for $0 \leq s \leq \frac{m_j}{2\beta_j}$. Now define a piecewise differentiable curve $\gamma(t)$, $0 \leq t \leq t_j + r$, by

$$\gamma(t) = \begin{cases} (x_1 + t, x_2, ..., x_n), & 0 \leq t \leq t_j \\ (x_1 + t_j, x_2..., x_j + \frac{1}{2}m_j(t - t_j), ..., x_n), & t_j \leq t \leq t_j + r \end{cases}.$$

For $r \leq \frac{1}{\beta_j}$ and $t_j \leq t \leq t_j + r$,

$$0 \leq \frac{1}{2}m_j(t - t_j) \leq \frac{r}{2}m_j \leq \frac{m_j}{2\beta_j},$$

and we conclude from (245) that

$$\gamma_j'(t) = \frac{1}{2}m_j \leq a_j(\gamma(t)).$$

Thus $\gamma(t)$ is subunit and

$$K_j(x, 2r) \geq |\gamma_j(t_j + r) - x_j| = \frac{r}{2}m_j \geq C_1 A_j(x, r) > C_0 A_j(x, 2r),$$

by doubling, with $\frac{C_1}{C_0}$ chosen to exceed the doubling constant. This yields the desired contradiction that $K(x, 2r)$ is not contained in $C_0 A(x, 2r)$. Therefore the second statement implies the third, and completes the proof of Proposition 104.

4. Reverse Hölder examples

In this subsection of the appendix we construct for every $0 < \varepsilon < 1$ a nonnegative function a_ε on the interval $I_0 = [-1, 1]$ that is Lipschitz continuous, doubling with doubling exponent $d = 3$, and that satisfies

$$a_\varepsilon \in RH_p, \quad p < \frac{1}{1 - \varepsilon},$$

$$a_\varepsilon \notin RH_\infty.$$

We can use such a weight as one of the a_j in Theorem 24, say $a_j(x) = a_\varepsilon(x_1)$, and by Proposition 104, the noninterference balls $A(x, r)$ are not equivalent to the subunit balls $K(x, r)$, thus providing an example of a homogeneous space structure *not* equivalent to the subunit balls of Fefferman and Phong that leads to a subellipticity theorem. Recall that w is doubling with exponent d and constant C if

$$(246) \quad \int_{x-r}^{x+r} w(t)\, dt \leq C\left(\frac{r}{s}\right)^d \int_{x-s}^{x+s} w(t)\, dt, \quad 0 < s < r, (x - r, x + r) \subset I_0.$$

The weight function

$$w_\varepsilon(t) = (1 + t^2)^{\frac{\varepsilon - 1}{2}}, \quad -\infty < t < \infty,$$

satisfies the doubling condition with exponent $d = 1$ and constant $C \approx \frac{1}{\varepsilon}$ for $\varepsilon > 0$, has bounded derivative if $\varepsilon \leq 2$, satisfies the RH_p condition for $p < \frac{1}{1-\varepsilon}$, and fails to satisfy the RH_∞ condition (consider the sequence of intervals $[0, R)$ as $R \to \infty$). Our goal is to construct a function $a_\varepsilon(t)$ with similar properties on the finite interval $|t| \leq 1$.

Note that both the doubling and RH_p conditions are invariant under dilations, translations and multiplication by constants. For $n = 1, 2, \ldots$, let

$$\alpha_n = \frac{1}{n!}, \quad \beta_n = n\alpha_n^2 = \frac{n}{(n!)^2},$$

and define

$$a_{n,\varepsilon}(t) = \beta_n w_\varepsilon\left(\frac{t-\alpha_n}{\beta_n}\right)$$

for fixed ε with $0 < \varepsilon < 1$. Then $a_{n,\varepsilon}(t)$ is a doubling weight with exponent $d = 1$ uniformly in n, i.e., its doubling constant C in (246) with $d = 1$ is independent of n (but dependent on ε of course). Let $\phi(t)$ be a smooth, nondecreasing function defined for $t \in [0,\infty)$ with $0 \le \phi(t) \le 1$, $\phi(t) = 0$ if $t \le 1/4$, $\phi(t) = 1$ if $t \ge 3/4$, and $\phi'(t) \le 3$. Now let $a_\varepsilon(t)$ be the even function defined for $0 \le t \le 1$ by

$$a_\varepsilon(t) = t^2 + \sum_{n=1}^\infty a_{n,\varepsilon}(t)\,\phi\left(\frac{t}{\alpha_n}\right).$$

Note that the weight $w(t) = t^2$ has doubling exponent $d = 3$.

We claim that on the interval $|t| \le 1$, $a_\varepsilon(t)$ is a Lipschitz continuous doubling weight with exponent $d = 3$ which satisfies the RH_p condition for $p < \frac{1}{1-\varepsilon}$, and fails to satisfy RH_∞.

Lipschitz Continuity. Let us first show that $a_\varepsilon(t)$ is Lipschitz continuous on $|t| \le 1$. Clearly, for $0 \le t \le 1$,

$$a_\varepsilon'(t) = 2t + \sum_{n=1}^\infty (\varepsilon-1)\frac{t-\alpha_n}{\beta_n}\left\{1+\left(\frac{t-\alpha_n}{\beta_n}\right)^2\right\}^{\frac{\varepsilon-3}{2}}\phi\left(\frac{t}{\alpha_n}\right)+\sum_{n=1}^\infty a_{n,\varepsilon}(t)\frac{1}{\alpha_n}\phi'\left(\frac{t}{\alpha_n}\right),$$

and so

$$|a_\varepsilon'(t)| \le 2 + \sum_{n=1}^\infty\left\{1+\left(\frac{t-\alpha_n}{\beta_n}\right)^2\right\}^{\frac{\varepsilon-2}{2}} + \sum_{n=1}^\infty 3\frac{\beta_n}{\alpha_n}.$$

Now use the fact that $|t-\alpha_n| \ge |\alpha_n|/4$ except for one value of n, in which case we will use the simple estimate

$$\left\{1+\left(\frac{t-\alpha_n}{\beta_n}\right)^2\right\}^{(\varepsilon-2)/2} \le 1.$$

We then obtain

$$|a_\varepsilon'(t)| \le 2 + 1 + \sum_{n=1}^\infty\left\{\frac{1}{4}(n-1)!\right\}^{\varepsilon-2} + 3\sum_1^\infty\frac{1}{(n-1)!} \le C$$

when $0 \le t \le 1$. It follows that $a_\varepsilon(t)$ is Lipschitz continuous on $|t| \le 1$.

Doubling with exponent $d = 3$. Now let us show that $a_\varepsilon(t)$ satisfies the doubling condition (246) with $d = 3$. Without loss of generality it is enough to show that for $0 < \varepsilon < 1$, there is a positive constant C_ε such that

$$\int_x^{x+r} a_\varepsilon(t)\,dt \le C_\varepsilon\left(\frac{r}{s}\right)^3\int_x^{x+s} a_\varepsilon(t)\,dt, \qquad 0 \le x < x+s < x+r \le 1,$$

Given $0 \leq x < x + s < x + r \leq 1$, let m be such that $\frac{\alpha_m}{4} < x + r \leq \frac{\alpha_{m-1}}{4}$. We now compute that

$$(247) \quad \int_x^{x+r} a_\varepsilon(t)\,dt \leq \int_x^{x+r} t^2\,dt + \sum_{n=m}^{\infty} \int_x^{x+r} a_{n,\varepsilon}(t)\,dt$$

$$= \frac{(x+r)^3 - x^3}{3} + \sum_{n=m}^{\infty} \beta_n \int_x^{x+r} w_\varepsilon\left(\frac{t-\alpha_n}{\beta_n}\right) dt$$

$$= r\left(x^2 + xr + \frac{1}{3}r^2\right) + \sum_{n=m}^{\infty} \beta_n^2 \int_{\frac{x-\alpha_n}{\beta_n}}^{\frac{x+r-\alpha_n}{\beta_n}} \left(1 + \tau^2\right)^{\frac{\varepsilon-1}{2}} d\tau$$

$$\leq r(x+r)^2 + C_\varepsilon \sum_{n=m}^{\infty} \beta_n^2 \left\{ \left(\frac{x+r-\alpha_n}{\beta_n}\right)^\varepsilon - \left(\frac{x-\alpha_n}{\beta_n}\right)^\varepsilon \right\}$$

$$\leq r(x+r)^2 + C_\varepsilon r \sum_{n=m}^{\infty} \beta_n^{2-\varepsilon} \leq C'_\varepsilon r(x+r)^2$$

since $\sum_{n=m}^{\infty} \beta_n^{2-\varepsilon} = \sum_{n=m}^{\infty} \left(n\alpha_n^2\right)^{2-\varepsilon} \leq C\left(m\alpha_m^2\right)^{2-\varepsilon} \leq C\alpha_m^2$ for m large, while

$$\int_x^{x+s} a_\varepsilon(t)\,dt \geq \int_x^{x+s} t^2\,dt = \frac{(x+s)^3 - x^3}{3} \geq \frac{1}{6} s(x+s)^2.$$

Combining these estimates yields

$$\int_x^{x+r} a_\varepsilon(t)\,dt \leq 6C'_\varepsilon \frac{r(x+r)^2}{s(x+s)^2} \int_x^{x+s} a_\varepsilon(t)\,dt \leq C_\varepsilon \left(\frac{r}{s}\right)^3 \int_x^{x+s} a_\varepsilon(t)\,dt,$$

as required.

Reverse Hölder of order $p < \frac{1}{1-\varepsilon}$. Suppose that $p < \frac{1}{1-\varepsilon}$. Then the weights t^2 and $a_{n,\varepsilon}(t)$ are in RH_p uniformly in n, say with constant $C_{p,\varepsilon}$. To show that

$$a_\varepsilon = t^2 + \sum_{n=1}^{\infty} a_{n,\varepsilon}(t)\phi(t/\alpha_n) \in RH_p,$$

fix an interval $I = [u,v] \subset [0,1]$. Let m be such that $\frac{\alpha_m}{4} < v \leq \frac{\alpha_{m-1}}{4}$ so that $a_\varepsilon(t) = t^2 + \sum_{n=m}^{\infty} a_{n,\varepsilon}(t)\phi(t/\alpha_n)$ for $t \in [u,v]$. Then

$$\left(\frac{1}{|I|}\int_I |a_\varepsilon|^p\right)^{\frac{1}{p}} \leq \left(\frac{1}{|I|}\int_I t^{2p}\right)^{\frac{1}{p}} + \sum_{n=m}^{\infty} \left(\frac{1}{|I|}\int_I |a_{n,\varepsilon}|^p\right)^{\frac{1}{p}}$$

$$\leq C_{p,\varepsilon}\left\{\frac{1}{|I|}\int_I t^2 + \sum_{n=m}^{\infty} \frac{1}{|I|}\int_I a_{n,\varepsilon}\right\}$$

$$= C_{p,\varepsilon}\frac{1}{|I|}\int_I a_\varepsilon.$$

Failure of RH_∞. Let $I_m = [0, \alpha_m]$. Then

$$\frac{1}{|I_m|}\int_{I_m} a_\varepsilon(t)\,dt = \frac{1}{\alpha_m}\int_0^{\alpha_m} a_\varepsilon(t)\,dt \leq C'_\varepsilon \alpha_m^2$$

by (247) with $x = 0$. But $a_\varepsilon(\alpha_m) \geq \beta_m w_\varepsilon(0)\phi(1) = m\alpha_m^2$, and combining estimates we obtain

$$\sup_{t \in I_m} a_\varepsilon(t) \geq a_\varepsilon(\alpha_m) \geq m\alpha_m^2 \geq \frac{m}{C'_\varepsilon}\frac{1}{|I_m|}\int_{I_m} a_\varepsilon(t)\,dt,$$

which shows that a_ε fails to satisfy RH_∞.

5. Product reverse Hölder

We will now show that (225) holds if either of the following two conditions is satisfied:

either each a_j satisfies the RH_∞ condition in each variable x_i, $i \neq 1, j$, uniformly in the other variables, i.e., there is a constant C such that for all $i, j = 2, \ldots, n$ with $i \neq j$ and all one-dimensional intervals I which contain x_i,
(248)
$$a_j(x_1, \ldots, x_n) \leq C \frac{1}{|I|} \int_I a_j(x_1, \ldots, x_{i-1}, z_i, x_{i+1}, \ldots, x_n) \, dz_i, \quad i, j \geq 2, i \neq j,$$

uniformly in x_k for $k \neq i$,

or the vector fields $\{a_j \frac{\partial}{\partial x_j}\}$ satisfy a strong non-interference condition relative to the boxes $CB(x, r)$ of the form

(249)
$$r \left(\sup_{z: |z_1 - x_1| \leq Cr; \, |z_k - x_k| \leq CB_k(x,r), k \geq 2} \left| \frac{\partial a_j}{\partial z_i}(z) \right| \right) B_i(x, r)$$
$$\leq \widetilde{C_0} B_j(x, r), \quad i, j \geq 2, i \neq j,$$

for all x and r and a suitably large geometric constant C. See Remark 19 for the interpretation of sup in (249). Note that we never consider the partial derivative $\partial / \partial z_1$ in (249). We will suppose that all $B_j(x, r) > 0$ if $r > 0$, which is of course implied by the flag condition 13.

For given j, x and r, the values of i in (249) can be restricted to those i which occur earlier than j in the ordering of the $B_j(x, r)$ since by Lemma 88 the condition automatically holds for the later i values. Of course the ordering generally varies with x and r. Note also that the RH_∞ condition (248) implies that there is a constant C such that if $i \geq 2$ and I is any one-dimensional interval which contains x_i, then

(250)
$$A_j(x, r) \leq \frac{C}{|I|} \int_I A_j((x_1, \ldots, x_{i-1}, z_i, x_{i+1}, \ldots, x_n), r) \, dz_i, \quad i \neq j.$$

To show (250) for example when $i = 2, j \neq 2$, note that by (248) and Fubini's theorem,

$$A_j(x, r) = \int_0^r a_j(x_1 + t, x_2, \ldots, x_n) \, dt \leq \int_0^r \left(\frac{C}{|I|} \int_I a_j(x_1 + t, z_2, x_3, \ldots, x_n) \, dz_2 \right) dt$$
$$= \frac{C}{|I|} \int_I A_j((x_1, z_2, x_3, \ldots, x_n), r) \, dz_2.$$

Similar estimates hold for Cartesian products of two or more intervals in different coordinates.

As will be apparent from the proof that (248) implies (225), instead of assuming (248) we actually only need a weak form of (250), "weak" in the sense that the average on the right side of (250) can be restricted to certain values of i with $i \geq 2$. In fact, if j occurs as $j = j_k$ in the ordering for the $B_\ell(x, r)$, we only need to assume (250) for $i = j_2, \ldots, j_{k-1}$ and not for $i = 1, j_k, \ldots, j_n$. We need the following two lemmas.

LEMMA 105. *If either (248) or (249) holds and $C \geq 1$, then*

$$B_j\left(x,r\right) \approx \max_{z:|z_1-x_1|\leq Cr;\ |z_i-x_i|\leq CB_i(x,r),i\geq 2} A_j\left(z,r\right)$$

with constants of equivalence that are independent of x and r.

PROOF. The proof is similar to that of Lemma 89. The fact that the left side of the conclusion is at most the right side is obvious from the definition of $B_j\left(x,r\right)$. To show the opposite inequality (with an appropriate constant factor), suppose we have the natural order and $x = 0$. Denoting $B_j\left(r\right) = B_j(0,r)$, we have by definition that

$$B_j\left(r\right) = \max_{|\xi_i|\leq B_i(r)} A_j\left((0,\xi_2,\ldots,\xi_{j-1},0,\ldots,0),r\right) \quad \text{if } j \geq 3 \text{ and} \quad B_2\left(r\right) = A_2(0,r).$$

Let $z = (z_1,\ldots,z_n)$ with $|z_1| \leq Cr$ and $|z_i| \leq CB_i\left(r\right)$ for $i \geq 2$. By doubling in the first variable,

$$A_j\left(z,r\right) \leq cA_j\left((0,z_2,\ldots,z_n),r\right).$$

If $j \geq 3$, the right side is at most

$$c\left|A_j\left((0,z_2,\ldots,z_n),r\right) - A_j\left((0,z_2^*,\ldots,z_{j-1}^*,0\ldots,0),r\right)\right|$$
$$+cA_j\left((0,z_2^*,\ldots,z_{j-1}^*,0,\ldots,0),r\right),$$

for $z_2^*,\ldots,z_{j-1}^*$ to be chosen. If $j \geq 3$ and $A_j(\cdot,r)$ satisfies (248), we pick $z_2^* = z_2,\ldots,z_{j-1}^* = z_{j-1}$, obtaining

$$\begin{aligned}
A_j\left(z,r\right) &\leq crC_L \sum_{i=j}^n |z_i| + cA_j\left((0,z_2,\ldots,z_{j-1},0,\ldots,0),r\right) \\
&\leq crC_L \sum_{i=j}^n CB_i\left(r\right) + cA_j\left((0,z_2,\ldots,z_{j-1},0,\ldots,0),r\right) \\
&\leq crC_L(n-1)C(1+rC_L)^{n-2}B_j\left(r\right) \\
&\quad +cA_j((0,z_2,\ldots,z_{j-1},0,\ldots,0),r)
\end{aligned}$$

by Lemma 88. By (248), the last term on the right, $cA_j\left((0,z_2,\ldots,z_{j-1},0,\ldots,0),r\right)$, is bounded by

$$\frac{c'}{\prod_2^{j-1} CB_i\left(r\right)} \int_{|\xi_i|\leq CB_i(r)} A_j\left((0,\xi_2,\ldots,\xi_{j-1},0,\ldots,0),r\right) d\xi_2\ldots d\xi_{j-1}$$

$$\leq \frac{c''}{\prod_2^{j-1} B_i\left(r\right)} \int_{|\xi_i|\leq B_i(r)} A_j\left((0,\xi_2,\ldots,\xi_{j-1},0,\ldots,0),r\right) d\xi_2\ldots d\xi_{j-1}$$

since RH_∞ implies doubling. The last expression is clearly at most

$$c'' \max_{|\xi_i|\leq B_i(r)} A_j\left((0,\xi_2,\ldots,\xi_{j-1},0,\ldots,0),r\right) = c''B_j\left(r\right),$$

which proves the lemma in this case.

If $j \geq 3$ and we instead assume the strong non-interference condition (249), we pick $z_2^* = \cdots = z_{j-1}^* = 0$ to get

$$A_j(z,r) \leq cr \sum_{i=2}^{n} \left(\sup_{|y_k| \leq |z_k|} \left| \frac{\partial a_j}{\partial y_i}(0, y_2, \ldots, y_n) \right| \right) |z_i| + cA_j(0,r)$$

$$\leq cr \sum_{i=2}^{n} \left(\sup_{|y_k| \leq |z_k|} \left| \frac{\partial a_j}{\partial y_i}(0, y_2, \ldots, y_n) \right| \right) CB_i(r) + cA_j(0,r)$$

$$\leq cr \sum_{i=2;\ i \neq j}^{n} \left(\sup_{|y_k| \leq |z_k|} \left| \frac{\partial a_j}{\partial y_i}(0, y_2, \ldots, y_n) \right| \right) CB_i(r) + crC_L CB_j(r) + cA_j(r)$$

$$\leq c(n-2)C\tilde{C}_0 B_j(r) + crC_L CB_j(r) + cB_j(r) \leq c'B_j(r),$$

and the proof is again complete (recall Remark 19 once more).

In the remaining case $j = 2$, we have the better estimate

$$\max_{|z_1| \leq Cr;\ |z_i| \leq CB_2(r), i \geq 2} A_2(z,r) \leq cB_2(r)$$

by (229) without requiring either (248) or (249). This completes the proof of the lemma.

We next show that either (248) or (249) implies a sort of RH_∞ condition for the $A_j(x,r)$. In fact, while each $B_j(x,r)$ is by definition the maximum of filled-out values of $A_j(\cdot, r)$, it turns out that $B_j(x,r)$ is no more than the corresponding integral average of $A_j(\cdot, r)$, assuming that (248) or (249) holds.

LEMMA 106. *Suppose that (248) or (249) holds and that the $B_j(x,r)$ are defined in the order $B_{j_k}(x,r)$, $k = 2, \ldots, n$. For each j_k, there are intervals*

$$I_j \subset [x_j - B_j(x,r), x_j + B_j(x,r)], \quad j = j_2, \ldots, j_{k-1},$$

depending on k, x and r, such that

$$|I_j| \approx B_j(x,r) \quad and \quad B_{j_k}(x,r) \approx \frac{1}{|I|} \int_I A_{j_k}(z,r) \, dz_{j_2} \cdots dz_{j_{k-1}} |_{z_i = x_i, i = 1, j_k, \ldots, j_n},$$

where $I = I_{j_2} \times \cdots \times I_{j_{k-1}}$. Also, if $C \geq 1$ and $J = J_{j_2} \times \cdots \times J_{j_{k-1}}$ for any intervals J_j which satisfy

$$I_j \subset J_j \subset [x_j - CB_j(x,r), x_j + CB_j(x,r)], \quad j = j_2, \ldots, j_{k-1},$$

then I may be replaced by J in the average above. If in addition $a_{j_k}(z)$ satisfies the doubling condition in each variable $z_i, i \neq j_k$ (which is automatic if (248) holds), then the requirement that $I_j \subset J_j$ can be dropped, i.e., we have

$$(251) \qquad B_{j_k}(x,r) \approx \frac{1}{|J|} \int_J A_{j_k}(z,r) \, dz_{j_2} \cdots dz_{j_{k-1}} |_{z_i = x_i, i = 1, j_k, \ldots, j_n}$$

for any $J = J_{j_2} \times \cdots \times J_{j_{k-1}}$ which satisfies

$$(252) \qquad J_j \subset [x_j - CB_j(x,r), x_j + CB_j(x,r)] \quad and \quad |J_j| \geq cB_j(x,r), \ c > 0.$$

PROOF. Assume first that (249) holds. When $k = 2$, the results follow from (229) even without assuming (249). For $k \geq 3$, we give the proof for the case $x = 0$ with the natural order. For $j \geq 3$, the inequality

$$B_j(r) \geq \frac{1}{|I|} \int_I A_j((0, z_2, \ldots, z_{j-1}, 0, \ldots, 0), r) \, dz_2 \cdots dz_{j-1}$$

is obvious from the definition of $B_j(r)$ for any $I = I_2 \times \cdots \times I_{j-1}$ with $I_i \subset [-B_i(r), B_i(r)]$. Now pick $\bar{z}_2, \ldots, \bar{z}_{j-1}$ with $|\bar{z}_i| \le B_i(r)$ and

$$B_j(r) = A_j((0, \bar{z}_2, \ldots, \bar{z}_{j-1}, 0 \ldots, 0), r).$$

For $i = 2, \ldots, j-1$, let I_i be an interval with $I_i \subset [-B_i(r), B_i(r)], \bar{z}_i \in I_i$, and $|I_i| = \varepsilon B_i(r)$ for ε to be chosen. Letting $I = I_2 \times \cdots \times I_{j-1}$, we have

$$\left| B_j(r) - \frac{1}{|I|} \int_I A_j((0, z_2, \ldots, z_{j-1}, 0, \ldots, 0), r) \, dz_2 \cdots dz_{j-1} \right|$$

$$\le \frac{1}{|I|} \int_I |A_j((0, \bar{z}_2, \ldots, \bar{z}_{j-1}, 0, \ldots, 0), r)$$

$$- A_j((0, z_2, \ldots, z_{j-1}, 0, \ldots, 0), r)| \, dz_2 \cdots dz_{j-1}$$

$$\le \frac{1}{|I|} \int_I r \sum_{i=2}^{j-1} \left(\sup_{|y_k| \le B_k(r)} \left| \frac{\partial a_j}{\partial y_i}(0, y_2, \ldots, y_n) \right| \right) |I_i| \, dz_2 \cdots dz_{j-1}$$

$$= r \sum_{i=2}^{j-1} \left(\sup_{|y_k| \le B_k(r)} \left| \frac{\partial a_j}{\partial y_i}(0, y_2, \ldots, y_n) \right| \right) \varepsilon B_i(r) \le (n-2)\varepsilon \tilde{C}_0 B_j(r)$$

by (249). The first part of the lemma now follows by picking ε with $(n-2)\varepsilon \tilde{C}_0 = 1/2$ and absorbing the last term.

To prove the second statement, still assuming that (249) holds, let $J = J_2 \times \cdots \times J_{j-1}$ for intervals J_i with

$$I_i \subset J_i \subset [-CB_i(r), CB_j(r)], \quad i = 2, \ldots, j-1.$$

Here C is fixed with $C \ge 1$. Lemma 105 gives

$$B_j(r) \ge c \frac{1}{|J|} \int_J A_j((0, z_2, \ldots, z_{j-1}, 0, \ldots, 0), r) \, dz_2 \cdots dz_{j-1}$$

for some $c > 0$, and the opposite inequality follows from the first statement of the lemma since $I_i \subset J_i$ and $|J_i|, |I_i| \approx B_i(r)$. Finally, in case $A_j((0, z_2, \ldots, z_{j-1}, 0, \ldots, 0), r)$ satisfies the doubling condition in $z_2, \ldots, z_{j-1}$, we have for any J of the type described in the third statement that

$$\frac{1}{|J|} \int_J A_j((0, z_2, \ldots, z_{j-1}, 0, \ldots, 0), r) \, dz_2 \cdots dz_{j-1} \approx$$

$$\frac{1}{\prod_2^{j-1} B_i(r)} \int_{|z_i| < B_i(r)} A_j((0, z_2, \ldots, z_{j-1}, 0, \ldots, 0), r) \, dz_2 \cdots dz_{j-1}.$$

Hence, the third statement follows from the second one, and the proof is complete in case (249) holds.

The proof is simpler in case the RH_∞ condition (248) holds. In fact, the conclusions then follow easily from Lemma 89, Lemma 105 and condition (248). The intervals I_j in the first statement can be taken to be $I_j = [x_j - B_j(x, r), x_j + B_j(x, r)]$. The doubling condition which is assumed in the last part of the lemma is automatically true by (248). This completes the proof of Lemma 106.

In Lemma 106, it is always possible to choose $J_j = [x_j - B_j(x,r), x_j + B_j(x,r)]$, obtaining as a corollary that

$$B_{j_k}(x,r) \approx \frac{1}{\prod_{j=j_2,\ldots,j_{k-1}} B_j(x,r)}$$
$$\times \int_{|z_j - x_j| \le B_j(x,r)} A_{j_k}(z,r)\, dz_{j_2} \ldots dz_{j_{k-1}} \,|_{z_i = x_i, i=1, j_k, \ldots, j_n}.$$

In this integral average, the fixed variables $z_1, z_{j_k}, \ldots, z_{j_n}$ that are chosen above to be $x_1, x_{j_k}, \ldots, x_{j_n}$, respectively, can instead be chosen to be any values $y_1, y_{j_k}, \ldots, y_{j_n}$ with $|y_1 - x_1| \le r$ and $|y_j - x_j| \le B_j(x,r), j = j_k, \ldots, j_n$, provided r is small. The proof is like the one given in the Remark after (229). It follows that we may also average over such values $y_1, y_j, \ldots, y_n$. In this way we obtain that if either (248) or (249) holds, then

$$(253) \qquad B_{j_k}(x,r) \approx \frac{1}{r \prod_2^n B_i(x,r)} \int_{|z_1 - x_1| \le r;\ |z_i - x_i| \le B_i(x,r)} A_{j_k}(z,r)\, dz_1 \ldots dz_n.$$

Finally, by combining (253) with Lemma 105, we immediately obtain the first of our two main results of this subsection:

PROPOSITION 107. *If either (248) or (249) holds, then so does (225).*

We now formulate a variant of Theorem 17 that uses the noninterference condition (249) in place of the reverse Hölder condition (248).

THEOREM 108. *Suppose for $1 \le j \le n$ that $a_j(x)$ is nonnegative and Lipschitz continuous on a domain $\Omega \subset \mathbb{R}^n$. Let $\mathcal{B} = \{B(x,r)\}$ denote the family of flag balls constructed in subsections 2.2.2 and 5.2.1 and suppose that the strong noninterference condition (249) holds. Assume as well that each $a_j(x)$ is reverse Hölder of infinite order in x_1 uniformly in $x_2, \ldots, x_n$, and satisfies the A_∞ condition in each variable x_i with $i \ne j$, uniformly in the remaining variables as in (234). Then the set $\mathcal{X}$ of vector fields $X_j = a_j \frac{\partial}{\partial x_j}$, $1 \le j \le n$, is subelliptic in Ω if and only if $\mathcal{X}$ satisfies the flag condition in Definition 13 in Ω. In the case that the flag condition holds in Ω, there is $Q \in [2,\infty)$ depending only on the Lipschitz and reverse Hölder constants of the a_j such that $\mathcal{X}$ is L^q-subelliptic in Ω for all $q > Q$.*

6. The noninterference conditions

We will now show that the noninterference condition (217) (or that in Definition 18) is a corollary of the *strong* noninterference condition (41): there are positive constants C_0 and r_0 such that for all x and all r with $0 < r \le r_0$,

$$(254) \qquad r \sup_{z \in A(x,r)} \left| \frac{\partial a_i}{\partial z_j}(z) \right| A_j(x,r) \le C_0 A_i(x,r), \qquad i, j \ge 2.$$

In case $n = 2$, condition (254) holds vacuously. In case $n > 2$ and all a_i are equal, it automatically holds because of the Lipschitz continuity of a_i in $x_2, \ldots, x_n$. In any case, (254) holds for $i = j$ by Lipschitz continuity.

LEMMA 109. *Suppose that each $a_i(x)$ satisfies the doubling condition in x_1 uniformly in $x_2, \ldots, x_n$, with doubling constant C_d, and is Lipschitz continuous in*

$x_2, \ldots, x_n$ *uniformly in* x_1. *If* (254) *holds, then* (217) (*or Definition* 18) *holds, i.e. there is a positive constant* C_c *and* r_0 *such that for* $0 < r \leq r_0$ *and* $2 \leq i \leq n$,

$$(255) \qquad C_c^{-1} A_i(x,r) \leq A_i(z,r) \leq C_c A_i(x,r), \qquad z \in A(x,r).$$

PROOF. For $\varepsilon > 0$, consider the "flattened" rectangle

$$A^\varepsilon(x,r) = (x_1 - r, x_1 + r) \times \prod_{i=2}^{n} (x_i - \varepsilon A_i(x,r), x_i + \varepsilon A_i(x,r)).$$

We begin by showing that (254) implies the following weaker form of (255), which requires comparability only over the flattened rectangle $A^\varepsilon(x,r)$; there are positive constants C_c, ε and r_0 such that for $0 < r \leq r_0$ and $2 \leq i \leq n$,

$$(256) \qquad C_c^{-1} A_i(x,r) \leq A_i(z,r) \leq C_c A_i(x,r), \qquad z \in A^\varepsilon(x,r).$$

To see this, fix $A^\varepsilon(x,r)$ with ε and r to be chosen in a moment. For simplicity, suppose that $x = 0$. If $z \in A^\varepsilon(0,r)$, then by doubling of a_i in the first variable,

$$A_i(z,r) \approx A_i((0, z_2, \ldots, z_n), r),$$

with constants of equivalence C_d and C_d^{-1}. Moreover,

$$
\begin{aligned}
|A_i((0, z_2, \ldots, z_n), r) - A_i(0,r)| &\leq \int_0^r |a_i(t, z_2, \ldots, z_n) - a_i(t, 0, \ldots, 0)| \, dt \\
&\leq \int_0^r \left(\sum_{j=2}^{n} \sup_{\xi \in A(0,r)} \left| \frac{\partial a_i}{\partial \xi_j}(\xi) \right| |z_j| \right) dt \\
&\leq r \sum_{j=2}^{n} \sup_{\xi \in A(0,r)} \left| \frac{\partial a_i}{\partial \xi_j}(\xi) \right| \varepsilon A_j(0,r), \\
&\leq \sum_{j=2}^{n} \varepsilon C_0 A_i(0,r) = (n-1) \varepsilon C_0 A_i(0,r),
\end{aligned}
$$

by (254) if $r \leq r_0$. If now ε is chosen so that $\varepsilon(n-1) C_0 = \frac{1}{2}$, we may absorb the last term to obtain

$$\frac{1}{2} A_i(0,r) \leq A_i((0, z_2, \ldots, z_n), r) \leq \frac{3}{2} A_i(0,r).$$

Combining estimates gives

$$\frac{1}{2C_d} A_i(0,r) \leq A_i(z,r) \leq \frac{3C_d}{2} A_i(0,r),$$

which is (256) as claimed. The proof of the lemma is now completed by the following remark.

REMARK 110. *The flattened comparability condition* (256) *implies* (255). *Indeed, suppose flattened rectangles* $A^\varepsilon(x,r)$ *and* $A^\varepsilon(y,r)$ *are adjacent, i.e.* $x_1 = y_1$, *and intersect, say*

$$z \in A^\varepsilon(x,r) \cap A^\varepsilon(y,r).$$

Then applying (256) *first to* $z \in A^\varepsilon(x,r)$, *and then to* $z \in A^\varepsilon(y,r)$, *we obtain*

$$(257) \qquad A_i(x,r) \approx A_i(z,r) \approx A_i(y,r), \qquad 2 \leq i \leq n.$$

There is an integer N, depending only on ε and the dimension n, with the following property. There are points $y^1, ..., y^N$ in $A(x,r)$ with $y_1^j = x_1$, $1 \le j \le N$, such that

(258) $$A(x,r) \subset \cup_{j=1}^N A^\varepsilon(y^j, r).$$

The full comparability condition (255) follows easily from (256), (257) and (258) since any two points u, v in $A(x,r)$ lie in flattened rectangles $A^\varepsilon(y^j, r), A^\varepsilon(y^k, r)$ respectively, which can be connected by a chain of adjacent flattened rectangles $A^\varepsilon(y^\ell, r)$ of length at most N.

We now demonstrate that the noninterference condition in Definition 18 is automatic if $A_i = A_j$ for $2 \le i, j \le n$.

LEMMA 111. *Suppose that each $a_i(x)$ satisfies the doubling condition in x_1 uniformly in $x_2, ..., x_n$, with doubling constant C_d, and is Lipschitz continuous in $x_2, ..., x_n$ uniformly in x_1. Suppose moreover that $a_1 = 1$ and $a_i = a_j$ for $2 \le i, j \le n$. Then*

$$C_c^{-1} A_j(x,r) \le A_j(z,r) \le C_c A_j(x,r), \quad z \in A(x,r),$$

for $2 \le j \le n$ and r small.

PROOF. For $2 \le j \le n$, let $a_j = a$ and $A_j = A$. Suppose $x = 0$ and consider the integral curve $\gamma(t)$ defined by $\gamma(0) = 0$ and $\gamma'(t) = (1, a(\gamma(t)), 0')$. Then

$$\gamma(r) = \left(r, \int_0^r a(\gamma(t)) \, dt, 0' \right),$$

and we have the estimate

$$\left| \int_0^r a(\gamma(t)) \, dt - A(0,r) \right| = \left| \int_0^r [a(\gamma(t)) - a(t,0,0')] \, dt \right|$$

$$\le \int_0^r \left| a\left(t, \int_0^t a(\gamma(s)) \, ds, 0' \right) - a(t,0,0') \right| dt$$

$$\le r \|\nabla' a\|_\infty \int_0^r a(\gamma(s)) \, ds,$$

where $\nabla' = \left(\frac{\partial}{\partial x_2}, ..., \frac{\partial}{\partial x_n} \right)$. Since a is Lipschitz in $x_2, ..., x_n$, we conclude that for r sufficiently small, depending only on $\|\nabla' a\|_\infty$,

$$\frac{1}{2} A(0,r) \le \int_0^r a(\gamma(t)) \, dt \le 2A(0,r).$$

Thus if $0 \le z_2 \le \frac{1}{2} A(0,r)$, there is $t \in [0,r]$ such that the point $z = (t, z_2, 0')$ lies on the image of the integral curve $\gamma(t)$, $0 \le t \le r$, and a similar argument shows that

$$\frac{1}{2} A(z,r) \le \int_0^r a(\gamma(t)) \, dt \le 2A(z,r).$$

The case $-\frac{1}{2} A(0,r) \le z_2 \le 0$ is similar and omitted. Using the doubling of $A(x,r)$ in r we conclude that

$$A(0,r) \approx A((0, z_2, 0'), r), \quad \text{for } |z_2| \le \frac{1}{2} A(0,r).$$

Repeating this argument with $x = (0, z_2, 0')$ and $|z_3| \le \frac{1}{2} A((0, z_2, 0'), r)$, we obtain

$$A(0,r) \approx A((0, z_2, z_3, 0'), r), \quad \text{for } |z_2|, |z_3| \le \frac{1}{2} A(0,r).$$

Continuing in this way, we finally obtain

(259) $A(0, r) \approx A(z, r), \quad for\ z \in A^{\frac{1}{2}}(x, r).$

This establishes the flattened comparability condition (256), and Remark 110 now completes the proof of the lemma.

7. Other notions of weak solution

We now discuss alternative definitions of a weak solution of the general divergence form equation (22),

(260) $\mathcal{L}u \equiv Lu + \mathbf{HR}u + \mathbf{S'G}u + Fu = f + \mathbf{T'g}.$

Recall that a classical weak solution to this equation is a function $u \in W^{1,2}(\Omega)$ such that (102) holds, i.e.

(261) $-\int (\nabla u)'\ B\nabla w + \int (\mathbf{HR}u)\,w + \int u\mathbf{GS}w + \int Fuw = \int fw + \int \mathbf{gT}w,$

for all test functions $w \in W_0^{1,2}(\Omega)$. As pointed out in Remark 51, we can make sense of this using the usual Sobolev embedding theorem, $W^{1,2}(\Omega) \subset L_{loc}^{q_n}(\Omega)$ ($q_n = \frac{2n}{n-2}$ if $n \geq 3$, $q_n < \infty$ if $n = 2$), for $u \in W^{1,2}(\Omega)$ and $f, F \in L_{loc}^{\frac{n}{2}}(\Omega)$ and $\mathbf{g}, \mathbf{G}, \mathbf{H} \in L_{loc}^n(\Omega)$. We can also consider other notions of weak solution to (260) by requiring that u belong to a Banach space B, and that w range over the dual Banach space B $'_0$, where the duality pairing is given by

$$\langle u, w \rangle = \int_\Omega u(x)\,w(x)\,dx + \int_\Omega (\nabla w)(x)'\ P(x)\,\nabla u(x)\,dx,$$

for some nonnegative semidefinite matrix $P(x) \succeq cB(x)$, where $B(x)$ is the coefficient matrix of L. The coefficients and data in (260) must then be in appropriate spaces depending on the Banach space B and the matrix $P(x)$.

For example, we may take B $= W_{\mathcal{Q}}^{1,2}(\Omega)$ and $P(x) = Q(x)$ where the quadratic form $\mathcal{Q}(x, \xi) = \xi'Q(x)\xi$ and $W_{\mathcal{Q}}^{1,2}(\Omega)$ denotes the completion of $Lip_1(\Omega)$ under the norm

$$\|w\|_{W_{\mathcal{Q}}^{1,2}(\Omega)} = \left\{ \int_\Omega \left(|w|^2 + \|\nabla w\|_{\mathcal{Q}}^2 \right) \right\}^{\frac{1}{2}}.$$

The integrals in (261) then make sense for $\xi'B(x)\xi \approx \mathcal{Q}$ as in (27) using the Sobolev embedding (15), together with the assumptions

$$\|f\|_{L_{loc}^{\frac{\sigma'}{2}}(\Omega)}, \|\mathbf{g}\|_{L_{loc}^{\sigma'}(\Omega)} < \infty,$$

and

$$\|F\|_{L_{loc}^{\frac{\sigma'}{2}}(\Omega)}, \|\mathbf{G}\|_{L_{loc}^{\sigma'}(\Omega)}, \|\mathbf{H}\|_{L_{loc}^{\sigma'}(\Omega)} < \infty.$$

Note that the weak solution u is permitted to be a much more general object here, namely a sequence of Lipschitz functions Cauchy in $W_{\mathcal{Q}}^{1,2}$ norm.

We close this subsection with a general comment, not to be taken too literally. There are analogues of our main theorems for weak solutions $u \in W_{\mathcal{Q}}^{1,2}(\Omega)$ in the sense discussed above, and we conjecture that with this notion of weak solution, and under the hypotheses of our main theorems, $\mathcal{Q}$ is both L^q-subelliptic relative to the homogeneous Dirichlet problems for the balls $B(x, r)$, and L^2-hypoelliptic relative to the homogeneous Neumann problems for the balls $B(x, r)$ (see subsection 1.1.1 for terminology).

8. Alternate methods of proof

We now discuss alternate methods of proof of some of the results in this paper.

First, it should be mentioned that both of our extensions, Theorems 17 and 20, of Hörmander's commutation theorem, can be proved without recourse to our generalization, Theorem 10, of the subellipticity theorem of Fefferman and Phong. Indeed, the proofs of the sharp versions, Theorems 23 and 24, use Theorem 8 with flag and noninterference balls, and do not rely on subunit balls at all.

Second, it is possible to use the method of Bombieri (Lemma 3 in [**31**]) to bridge the gap at $\beta = 0$ in the Moser iteration in section 3, rather than using bounded mean oscillation and the John-Nirenberg theorem. The idea is to use the arguments in Lemmas 83 and 84 to create "accumulating sequences of Lipschitz cutoff functions" in arbitrarily thin annuli of the form $A(x,s) \setminus A(x,t)$ and $\frac{t}{r}B(x,r) \setminus \frac{s}{r}B(x,r)$ respectively, for $\frac{1}{2}r < s < t \leq r$. The details are left to the interested reader.

Third, it is possible to prove the basic Lemma 79 for the noninterference balls $A(x,r)$ by adapting Franchi's method [**8**] of varying the driving parameters $u = (u_2, ..., u_n)$ for *fixed* initial point x in the integral curves $\gamma_u(x,t)$ in subsection 4.4. However, in dealing with the flag balls $B(x,r)$ it is convenient, if not necessary, to instead vary the initial points x for an appropriately *fixed* driving parameter u in the integral curves $\gamma_u(x,t)$.

Bibliography

[1] S. CAMPANATO, Equazioni ellittiche del II ordine e spazi $\mathcal{L}^{(2,\lambda)}$, *Ann. Mat. Pura Appl. (4)* **69** (1965), 321-381.

[2] S. CAMPANATO, *Sistemi ellittici in forma divergenza. Regolarità all' interno*, Quaderni. Pisa: Scuola Normale Superiore 1980.

[3] S. CHANILLO AND R. L. WHEEDEN, Harnack's inequality and mean-value estimates for solutions of degenerate elliptic equations, *Comm. P. D. E.* **11** (1986), 1111–1134.

[4] M. CHRIST, A $T(b)$ theorem with remarks on analytic capacity and the Cauchy integral, *Colloq. Math.* **61** (1990), 601-628.

[5] R. COIFMAN AND G. WEISS, Analyse harmonique non-commutative sur certain espaces homogènes, *Lecture Notes in Math.* **242**, Springer-Verlag, New York, Berlin 1971.

[6] E. B. FABES, C. E. KENIG AND R. P. SERAPIONI, The local regularity of solutions of degenerate elliptic equations, *Comm. P. D. E.* **7** (1982), 77–116.

[7] C. FEFFERMAN AND D. H. PHONG, Subelliptic eigenvalue problems, *Conf. in Honor of A. Zygmund,* Wadsworth Math. Series 1981.

[8] B. FRANCHI, Weighted Sobolev-Poincaré inequalities and pointwise estimates for a class of degenerate elliptic equations, *Trans. Amer. Math. Soc.* **327** (1991), 125-158.

[9] B. FRANCHI AND E. LANCONELLI, Hölder regularity theorem for a class of linear nonuniformly elliptic operators with measurable coefficients, *Annali Scuola Normale Sup. Pisa 4,* **10** (1983), 523-541.

[10] B. FRANCHI, C. PÉREZ AND R. L. WHEEDEN, Self-improving properties of John–Nirenberg and Poincaré inequalities on spaces of homogeneous type, *J. Functional Analysis* **153** (1998), 108–146.

[11] B. FRANCHI, R. SERAPIONI AND F. SERRA CASSANO, Approximation and imbedding theorems for weighted Sobolev spaces associated with Lipschitz continuous vector fields, *Boll. Un. Mat. Ital. B (7)* **11** (1997), no. 1, 83–117.

[12] N. GAROFALO AND D. M. NHIEU, Lipschitz continuity, global smooth approximations and extension theorems for Sobolev functions in Carnot-Carathéodory spaces, *J. Anal. Math.* **74** (1998), 67–97.

[13] M. GIAQUINTA, Multiple integrals in the calculus of variations and nonlinear elliptic systems, *Annals of Math. Studies* **105**, Princeton Univ. Press, New Jersey 1983.

[14] D. GILBARG AND N. TRUDINGER, *Elliptic Partial Differential Equations of Second Order*, Springer-Verlag, revised 3rd printing, 1998.

[15] P. GUAN, Regularity of a class of quasilinear degenerate elliptic equations, *Advances in Mathematics* **132** (1997), 24-45.

[16] P. GUAN, Quasilinear degenerate elliptic equations in divergence form, *Contemporary Mathematics* **205** (1997), 93-99.

[17] P. GUAN AND E. SAWYER, Regularity estimates for the oblique derivative problem, *Annals of Mathematics* **137** (1993), 1-71.

[18] C. E. GUTIÉRREZ, The Monge-Ampère equation, *Progress in Nonlinear Differential Equations and Their Applications*, volume **44**, Birkhauser 2002.

[19] C. E. GUTIÉRREZ AND E. LANCONELLI, Maximum principle, non-homogeneous Harnack inequality, and Liouville theorems for X-elliptic operators, *preprint*, 1-29.

[20] G. H. HARDY AND J. E. LITTLEWOOD, Some properties of conjugate functions, *J. Reine Angew. Math.* **167** (1932), 405–423.

[21] L. HORMANDER, Hypoelliptic second order differential equations, *Acta. Math.* **119** (1967), 141-171.

[22] D. JERISON, The Poincaré inequality for vector fields satisfying the Hörmander condition, *Duke Math. J.* **53** (1986), No. 2, 503-523.

[23] F. JOHN AND L. NIRENBERG, On functions of bounded mean oscillation, *Comm. Pure Appl. Math.* **14** (1961), 415-426.

[24] E. LANCONELLI AND D. MORBIDELLI, On the Poincaré inequality for vector fields, *Ark. Mat.* **38** (2000), 327-342.

[25] G. LU, Weighted Poincaré and Sobolev inequalities for vector fields satisfying Hörmander's condition and applications, *Revista Mat. Iberoamericana* **8** (1992), 367-439.

[26] G. LU, The sharp Poincaré inequality for free vector fields: an endpoint result, *Revista Mat. Iberoamericana* **10** (1994), 453-466.

[27] G. LU AND R. L. WHEEDEN, High order representation formulas and embedding theorems on stratified groups and generalizations, *Studia Math.* **142** (2000), 101-133.

[28] A. MONTANARI AND D. MORBIDELLI, Balls defined by nonsmooth vector fields and the Poincaré inequality, *preprint*, 1-18.

[29] C. B. MORREY JR., Second order elliptic systems of differential equations, In: *Annals of Math. Studies* **33**, 101-159, Princeton Univ. Press, New Jersey 1954.

[30] J. MOSER, On Harnack's theorem for elliptic differential equations, *Comm. Pure Appl. Math.* **14** (1961), 577-591.

[31] J. MOSER, On a pointwise estimate for parabolic differential equations, *Comm. Pure Appl. Math.* **24** (1971), 727–740.

[32] A. NAGEL, F. RICCI AND E. M. STEIN, Singular integrals with flag kernels and analysis on quadratic CR manifolds, *J. Functional Analysis* **181** (2001), 29–118.

[33] A. NAGEL, E. M. STEIN AND S. WAINGER, Balls and metrics defined by vector fields I: Basic properties, *Acta. Math.* **155** (1985), 103-147.

[34] C. RIOS, E. SAWYER AND R. L. WHEEDEN, A higher dimensional partial Legendre transform, and regularity of degenerate Monge-Ampère equations, *Advances in Math.* **193** (2005), 373–415.

[35] C. RIOS, E. SAWYER AND R. L. WHEEDEN, Regularity of degenerate quasilinear equations, *preprint*.

[36] L. ROTHSCHILD AND E. M. STEIN, Hypoelliptic differential operators and nilpotent groups, *Acta. Math.* **137** (1976), 247-320.

[37] W. RUDIN, *Principles of Real Analysis*, Wiley.

[38] E. SAWYER AND R. L. WHEEDEN, Regularity of degenerate Monge-Ampère and prescribed Gaussian curvature equations in two dimensions, *preprint*, available at http://www.math.mcmaster.ca/ sawyer

[39] E. SAWYER AND R. L. WHEEDEN, Regularity of degenerate Monge-Ampère and prescribed Gaussian curvature equations in two dimensions, *Potential Analysis*, to appear.

[40] E. SAWYER AND R. L. WHEEDEN, A priori estimates for quasilinear equations related to the Monge-Ampère equation in two dimensions, *J. d'Analyse Math.*, to appear.

[41] E. SAWYER AND R. L. WHEEDEN, Weighted inequalities for fractional integrals on Euclidean and homogeneous spaces, *Amer. J. Math.* **114** (1992), 813-874.

[42] E. M. STEIN, *Harmonic Analysis*, Princeton U. Press, 1993.

[43] J.-O. STROMBERG AND A. TORCHINSKY, Weighted Hardy Spaces, *Lecture Notes in Math.* **1381** (1989).

[44] M. TAYLOR, Partial Differential Equations, vol. 3, New York, Springer, 1996.

[45] C.-J. XU, Opérateurs sou-elliptiques et régularité des solitions aux dérivées partielles non-linéaires du second ordre en deux variables, *Comm. Partial Differential Equations* **11** (1986), 1575-1603.

[46] C.-J. XU AND C. ZUILY, Higher interior regularity for quasilinear subelliptic systems, *Prépublications Univ. Paris-Sud Math.* **425** (1995), 1-24.

Glossary of Terms and Symbols

Editorial Information

To be published in the *Memoirs*, a paper must be correct, new, nontrivial, and significant. Further, it must be well written and of interest to a substantial number of mathematicians. Piecemeal results, such as an inconclusive step toward an unproved major theorem or a minor variation on a known result, are in general not acceptable for publication. Papers appearing in *Memoirs* are generally at least 80 and not more than 200 published pages in length. Papers less than 80 or more than 200 published pages require the approval of the Managing Editor of the Transactions/Memoirs Editorial Board.

As of November 30, 2005, the backlog for this journal was approximately 15 volumes. This estimate is the result of dividing the number of manuscripts for this journal in the Providence office that have not yet gone to the printer on the above date by the average number of monographs per volume over the previous twelve months, reduced by the number of volumes published in four months (the time necessary for preparing a volume for the printer). (There are 6 volumes per year, each containing at least 4 numbers.)

A Consent to Publish and Copyright Agreement is required before a paper will be published in the *Memoirs*. After a paper is accepted for publication, the Providence office will send a Consent to Publish and Copyright Agreement to all authors of the paper. By submitting a paper to the *Memoirs*, authors certify that the results have not been submitted to nor are they under consideration for publication by another journal, conference proceedings, or similar publication.

Information for Authors

Memoirs are printed from camera copy fully prepared by the author. This means that the finished book will look exactly like the copy submitted.

The paper must contain a *descriptive title* and an *abstract* that summarizes the article in language suitable for workers in the general field (algebra, analysis, etc.). The *descriptive title* should be short, but informative; useless or vague phrases such as "some remarks about" or "concerning" should be avoided. The *abstract* should be at least one complete sentence, and at most 300 words. Included with the footnotes to the paper should be the 2000 *Mathematics Subject Classification* representing the primary and secondary subjects of the article. The classifications are accessible from www.ams.org/msc/. The list of classifications is also available in print starting with the 1999 annual index of *Mathematical Reviews*. The Mathematics Subject Classification footnote may be followed by a list of *key words and phrases* describing the subject matter of the article and taken from it. Journal abbreviations used in bibliographies are listed in the latest *Mathematical Reviews* annual index. The series abbreviations are also accessible from www.ams.org/publications/. To help in preparing and verifying references, the AMS offers MR Lookup, a Reference Tool for Linking, at www.ams.org/mrlookup/. When the manuscript is submitted, authors should supply the editor with electronic addresses if available. These will be printed after the postal address at the end of the article.

Electronically prepared manuscripts. The AMS encourages electronically prepared manuscripts, with a strong preference for $\mathcal{AMS}$-LaTeX. To this end, the Society has prepared $\mathcal{AMS}$-LaTeX author packages for each AMS publication. Author packages include instructions for preparing electronic manuscripts, the *AMS Author Handbook*, samples, and a style file that generates the particular design specifications of that publication series. Though $\mathcal{AMS}$-LaTeX is the highly preferred format of TeX, author packages are also available in $\mathcal{AMS}$-TeX.

Authors may retrieve an author package from e-MATH starting from www.ams.org/tex/ or via FTP to ftp.ams.org (login as anonymous, enter username as password, and type cd pub/author-info). The *AMS Author Handbook* and the *Instruction Manual* are available in PDF format following the author packages link from www.ams.org/tex/. The author package can be obtained free of charge by sending email to pub@ams.org (Internet) or from the Publication Division, American Mathematical Society, 201 Charles St., Providence, RI 02904, USA. When requesting an author package, please specify $\mathcal{AMS}$-LaTeX or $\mathcal{AMS}$-TeX, Macintosh or IBM (3.5) format, and the publication in which your paper will appear. Please be sure to include your complete mailing address.

Sending electronic files. After acceptance, the source file(s) should be sent to the Providence office (this includes any TeX source file, any graphics files, and the DVI or PostScript file).

Before sending the source file, be sure you have proofread your paper carefully. The files you send must be the EXACT files used to generate the proof copy that was accepted for publication. For all publications, authors are required to send a printed copy of their paper, which exactly matches the copy approved for publication, along with any graphics that will appear in the paper.

TeX files may be submitted by email, FTP, or on diskette. The DVI file(s) and PostScript files should be submitted only by FTP or on diskette unless they are encoded properly to submit through email. (DVI files are binary and PostScript files tend to be very large.)

Electronically prepared manuscripts can be sent via email to pub-submit@ams.org (Internet). The subject line of the message should include the publication code to identify it as a Memoir. TeX source files, DVI files, and PostScript files can be transferred over the Internet by FTP to the Internet node e-math.ams.org (130.44.1.100).

Electronic graphics. Comprehensive instructions on preparing graphics are available at www.ams.org/jourhtml/graphics.html. A few of the major requirements are given here.

Submit files for graphics as EPS (Encapsulated PostScript) files. This includes graphics originated via a graphics application as well as scanned photographs or other computer-generated images. If this is not possible, TIFF files are acceptable as long as they can be opened in Adobe Photoshop or Illustrator. No matter what method was used to produce the graphic, it is necessary to provide a paper copy to the AMS.

Authors using graphics packages for the creation of electronic art should also avoid the use of any lines thinner than 0.5 points in width. Many graphics packages allow the user to specify a "hairline" for a very thin line. Hairlines often look acceptable when proofed on a typical laser printer. However, when produced on a high-resolution laser imagesetter, hairlines become nearly invisible and will be lost entirely in the final printing process.

Screens should be set to values between 15% and 85%. Screens which fall outside of this range are too light or too dark to print correctly. Variations of screens within a graphic should be no less than 10%.

Inquiries. Any inquiries concerning a paper that has been accepted for publication should be sent directly to the Electronic Prepress Department, American Mathematical Society, 201 Charles St., Providence, RI 02904, USA.

Titles in This Series

TITLES IN THIS SERIES

For a complete list of titles in this series, visit the
AMS Bookstore at **www.ams.org/bookstore/**.